EARLY GREEK ASTRONOMY
to Aristotle

ASPECTS OF GREEK AND ROMAN LIFE

General Editor: Professor H. H. Scullard

EARLY GREEK ASTRONOMY
to Aristotle

D. R. Dicks

CORNELL UNIVERSITY PRESS
ITHACA, NEW YORK

Standard Book Number 8014-0561-0

Library of Congress Catalog Card Number 76-109335

PRINTED IN ENGLAND

CONTENTS

PREFACE

THE ACHIEVEMENTS of the Greeks in the field of astronomy must rank among the most enduring which even that gifted race has bequeathed to posterity. Every time we make use of a celestial globe, a planisphere, a star map, or a planetarium, we use concepts for which we are indebted to the ancient Greek astronomers; for it was they (and they alone of the peoples of antiquity) who conceived and developed the essential, theoretical basis of all such astronomical tools, i.e. the celestial sphere as a mathematical extension in space of the terrestrial sphere, with the observer envisaged as being at the centre of the cosmos.

Yet only one book dealing exclusively with ancient Greek astronomy has appeared in English since T. L. Heath's *Aristarchus of Samos* (Oxford, 1913), namely, the same scholar's much slighter *Greek Astronomy* (Dent, London, 1932), a selection of translations preceded by a 'potted' history of astronomy, mostly adapted from his larger work but with a few additional pages to bring the story down to Ptolemy (who receives eight lines!) and some unreliable material on Egyptian and Babylonian astronomy (not touched on in the earlier work). Heath's *Aristarchus* has held the field for over fifty years, just as his admirable *History of Greek Mathematics* (2 vols., Oxford, 1921) has become a classic in that sphere. Meanwhile, our knowledge of the history of science (particularly early science) has increased considerably. We know much more about Egyptian and Babylonian astronomical methods. We can assess to some extent how much the Greeks were influenced by or borrowed from these sources (although when and how such interaction took place are very debatable questions – see pp. 165*ff*.), and we are perhaps less inclined to accept uncritically the accounts of the doxographers and commentators of later antiquity who attribute comparatively advanced astronomical concepts to the early Greek

thinkers in defiance of all probability. Also our understanding of
Plato's astronomical ideas has been substantially enlarged by
F. M. Cornford's brilliant work on the *Timaeus* (*Plato's Cosmology*,
London, 1937).

The present volume attempts to take into account at least some
of the work done since the publication of Heath's *Aristarchus*. I
have been concerned to trace the development of *astronomy* as a
science, rather than that of cosmology or cosmogony, and so the
cosmological fantasies of the Pre-Socratics receive shorter shrift
than some scholars may think such theories merit. I make no
apology for this; elsewhere I have pointed out the fallacies
attendant on an overestimation of the scientific content of Pre-
Socratic thought (*CQ* 9, 1959, pp. 294–309; *JHS* 86, 1966, pp.
26–40), and the chief criticism that may be directed against the
earlier part of Heath's book is that he tends to treat as of equal
value all the evidence that has come down to us via the scrapbooks
and compilers of later times concerning the astronomical ideas of
the Pre-Socratics, and, without considering the implications, to
accept accounts as trustworthy that are only implausible guesses,
contradictory in themselves and anachronistic in the context.

Some justification is perhaps needed for the fact that this volume
ends with Aristotle, thus covering a shorter span than Heath, who
naturally took Aristarchus (some two generations after Aristotle)
as his terminal point. The main reason is that Plato (whose importance in the development of Greek astronomy is commonly underestimated) and Eudoxus are treated here at greater length (Chapters
V and VI), and more attention is paid to such topics as the origin
of the constellations (pp. 159ff.) and the influence of Babylonian
astronomy on that of the Greeks (pp. 163ff.). Aristotle makes a
particularly appropriate point at which to divide a history of
Greek astronomy, since after him mathematical astronomy under-
goes a relatively rapid development along lines that not only
contrast strongly with pre-Aristotelian speculative thought, but
also are substantially different from those laid down by Eudoxus
in Aristotle's own lifetime. It is, too, in the early history of astronomy, where the reliable evidence is scanty and its interpretation
uncertain, that the greatest number of erroneous notions and inept

judgments is commonly found; commentators (both ancient and modern) have been all too prone to read into early scientific thought ideas which are only appropriate to a later stage of development. It is hoped to follow the present volume with another dealing with later Greek astronomy, which will include some discussion of the influence of astrological doctrines on astronomical thought; traces of such doctrines can be found in Greek writings dating from the end of the fifth century BC, but astrology does not become influential in Greek thought until Hellenistic times (see my *Geographical Fragments of Hipparchus*, London, 1960, pp. 12*ff.*; *cf. Hermes* 91, 1963, pp. 60*ff.*), and so belongs properly to the second volume.

I am grateful to Professor H. H. Scullard, the General Editor of this series, for first suggesting that I should contribute an account of Greek astronomy, and to him and the publishers for agreeing that it should be in two volumes. Its preparation has been greatly facilitated by a year's leave of absence, generously granted by the authorities of Bedford College, to take up a Visiting Professorship at Princeton University for the first semester in 1966 and to work at the Institute for Advanced Study, Princeton, during the second semester. I acknowledge with deep gratitude the many kindnesses of Professor Marshall Claggett of the above Institute and of Professor Charles Gillispie of the Program in History and Philosophy of Science, Princeton University, who between them made all the necessary arrangements for my most enjoyable stay. I should also like to record my great appreciation of the work done by the members of the graduate seminar on Greek Astronomy which I was privileged to give; I learnt much from them, and was greatly impressed by their enthusiastic interest in and application to a somewhat esoteric subject – especially since not a single one of them was a classics graduate. Finally, although it may be invidious to single out one name when all the members of the above Program (and indeed of the whole Department of History) did so much so willingly to help me in different ways, I wish to tender my special thanks to Dr Michael Mahoney for his readiness to give unstintingly of his time and knowledge in both professional discussion and extra-curricular activities.

CHAPTER I

GENERAL PRINCIPLES[1]

ALL ANCIENT ASTRONOMY was based entirely on naked-eye observation (as, indeed, was all astronomy up to the end of the sixteenth century). This simple fact is often overlooked. Modern commentators tend to underestimate the difficulties of such observation and to discuss the results as though the telescope, the micrometer screw, and the vernier scale were normal pieces of equipment in antiquity. In reality, of course, none of these aids existed and without them it is by no means easy to make measurements of even a low standard of accuracy with the naked eye alone; intelligent application, assiduous practice in estimating changes of position in the night sky, and above all an excellent memory combined with access to similar observations recorded over as long a period as possible, are the minimum requirements to produce any meaningful results. To distinguish, for example, between a planet and a fixed star (easy enough with the lowest-powered binocular or telescope) is not at all a simple matter when one gazes on the myriad points of light in the heavens, especially when the atmosphere is clear (as might be expected in ancient Greece) and the phenomenon of 'twinkling' is less in evidence; to differentiate between the planets themselves, and estimate their periods, must have taken centuries of observation. We shall see later that the results obtained by even such a competent observer as Eudoxus (whose 'observatory' – probably nothing more than a flat-topped building – was still pointed out in the time of Strabo; see pp. 151f.) were far from accurate, and were reflected in the inadequacy of his solar theory in particular. There is practically no evidence as to what instruments were used to make astronomical observations before the third century BC.[2] It is a fair inference that measurements of the shadow cast by the gnomon, i.e. any vertical pointer on a horizontal base – probably introduced from Babylonia (see pp. 165f.) in the fifth century – were used to determine the solstices and the time of midday (see below), and it is likely that a simple rotating globe with the chief circles of the celestial sphere and some of the better-known constellations marked on it formed part of an astronomer's equipment from at least the time of Eudoxus onwards; but what sighting instrument was used (if any) is wholly unknown. It is not improbable that angular distances were estimated simply by means of the hand stretched out at arm's length in front of the eye; a possible reminiscence of such a usage may be found in the use of the word δάκτυλος ('finger' or 'finger's breadth') in later astronomy to express the degrees of totality of an eclipse, twelve 'fingers' defining a total eclipse (cf. Cleom., *Cycl. Theor.* ii, 3).

Now for naked-eye astronomy the significant objects observable are sun, moon, five planets only (Mercury, Venus, Mars, Jupiter, and Saturn), several thousand stars (of which only about one thousand, which can be classified in readily identifiable groupings, i.e. the forty-eight ancient constellations, were of real importance), plus various occasional phenomena such as eclipses, occultations of stars by the moon, comets, and shooting stars. All these are treated, of course, from the standpoint of a geocentric universe with the earth fixed firmly at the centre of the cosmos and the other celestial bodies moving round it in various orbits comprising circles or combinations of circles. As we shall see, there were suggestions from time to time in antiquity that this picture of the universe was not necessarily the only one that would account for the observed phenomena; but in the main, and certainly so far as concerns fully developed systems, the Greek astronomers, like all their successors up to Copernicus, were content with the geocentric model. It is from this point of view, which accepts as established fact that sun and stars move across the heavens and round the stationary earth once every twenty-four hours, just as they appear to do, that the following description is given of the basic astronomical phenomena with which the Greeks were concerned.[3]

Let us consider first the 'fixed' stars (τὰ ἀπλανῆ ἄστρα), so called to distinguish them from the sun, moon, and 'wandering' stars,

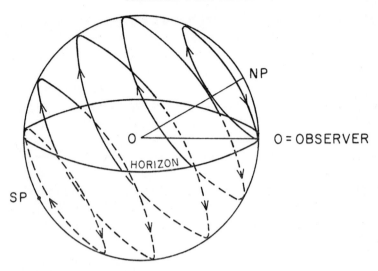

Fig. 1. Stellar risings and settings

i.e. the planets (οἱ πλάνητες); by their very number and the
regularity of their courses the fixed stars provide a striking
phenomenon for any nocturnal observer. If one observes the stars
night after night over a period of years one discovers three things
about them: (*a*) they all share in a continuous and uniform
wheeling motion over the sky in the general direction of east to
west – and the orbit of each star is curved; (*b*) whereas some of the
stars have large orbits, pass directly overhead, and are visible
throughout most of the night before setting below the western
horizon, while others have smaller orbits and can only be seen for
a short time, there is another group of constantly visible stars
which never rise or set and which seem to circle a particular point
in the sky; and (*c*) different stars are prominent at night at different
seasons of the year, but the same stars appear regularly at the same
places in the same seasons in successive years. Fig. 1 shows
schematically the state of affairs as seen in the northern hemisphere.
Here NP is the north pole and SP the south pole of the heavens; the
complete, continuous-line circle represents the circumpolar stars

that never dip below the horizon, while the complete, dotted-line circle represents the stars that are never seen above the horizon at this particular latitude. The reason why the stars of a summer night are different from those of a winter night is, of course, that the sidereal day (i.e. the time taken for one complete revolution of the heavens) is about 4 minutes shorter than the solar day of 24 hours, owing to the fact that the sun itself is moving steadily relative to the stars (see below). This means that a star that rises, say, 1 hour after sunset on a given night, will rise 4 minutes earlier on successive nights until eventually its rising will not be visible because the sun will not yet have set sufficiently below the horizon to enable the star to be seen in the still bright sky – when it *does* get dark enough, the star will already be well up in the night sky. Similarly, of course, at the other end of the night; if a star is seen rising just before the dawn (and is only just visible before the sun's rays outshine it), on succeeding nights it will rise earlier and earlier, its period of visibility reaching a maximum and then decreasing, until its rising is swallowed up by the sun's previous setting, and so can no longer be observed.

It was observation of these differences in the rising and setting of stars that enabled the ancients, layman and astronomer alike, to use the appearance of the heavens as a gigantic clock wherewith to measure the alternation of the seasons. Eight different stellar risings and settings (in relation to the sun) were classified;[4] four of these, called the true risings and settings, when the star in question rises or sets *exactly* with the sun, play no part in practical astronomy, because, of course, they are not observable. The visible risings and settings may be tabulated as follows:

Description	Characteristics	Technical Name
visible morning rising (vMR)	first visible appearance on eastern horizon before sunrise – the star is just sufficiently in advance of the sun to be visible for a moment – on subsequent nights the length of visibility progressively increases	heliacal rising
visible evening rising (vER)	last visible rising in the evening just after sunset – the next night the star will have risen on the eastern horizon while it is still too light to be seen – visible during the night	achronycal rising
visible morning setting (vMS)	first visible setting just before sunrise – the previous night the star had not yet reached the western horizon before sunrise – visible during the night	cosmical setting
visible evening setting (vES)	last visible setting just after sunset – the next night the star will have reached the western horizon while still too light to be seen – above horizon during the day	heliacal setting

For a period between the heliacal setting and the subsequent heliacal rising the star remains invisible at night; at other times the star is visible for various periods depending on the time of year, the position of the star itself relative to the sun, and the latitude of the observer.

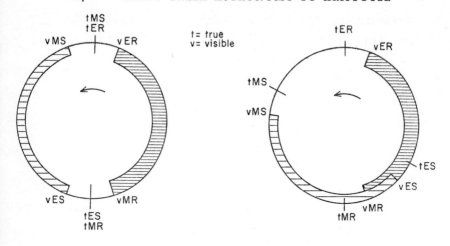

Figs 2, 3. *Visibility of stars on the ecliptic (left) and visibility of stars north of the ecliptic (right)*

A glance at Figs. 2 and 3[5] may clarify the matter. Fig. 2 shows the simplest case, of a star actually on the ecliptic (the ecliptic is the mathematical line on the celestial sphere that marks the annual path of the sun round the earth – see below), and the direction of movement is given by the arrow. The whole circle is one year; the shaded portions show the visibility of the star at setting and rising, and the parts in between are when the star is invisible. Fig. 3 shows the position for stars north of the ecliptic. Here it will be seen that the shaded portions overlap; this means that when the sun is located on the arc vMR to vES a star *can* (but this does not *always* happen – it depends on the position of the star) actually be seen *twice* in the same night. It is visible at the beginning of the night, it sets, remains invisible for some hours, and then can be seen rising again at the end of the night. A similar figure could be drawn for stars south of the ecliptic. The latitude of the observer is, of course, crucial, because for each particular horizon the stars observable and the times of rising and setting are different from any other. This is obvious from Fig. 1: tilt the horizon to form a different angle with the north-south axis (i.e. change the latitude) and it will intersect the stellar paths at different places. Very roughly, the more northerly the latitude, the earlier the stars north of the equator will rise and the later they will set (having a greater part of their circles above the horizon), while the more southerly the latitude, the earlier they will set and the later they will rise.

Now if this were the whole story, one might expect that the appearance of the night sky at a particular latitude would be the same now as it was 2,500 years ago. This, however, is not the case, owing chiefly to a phenomenon known as the precession of the equinoxes which results in a slow but continuous displacement westwards of the points where the ecliptic intersects the equator (i.e. the equinoxes – see below and *cf.* Fig. 4), so that every year the equinoctial points (and hence the solstitial points as well) shift about 50″ of arc along the ecliptic, completing the whole circuit of it in a period of some 25,800 years. This is because the earth's axis does not remain constant relative to the plane of the ecliptic, but exhibits a continuous tilting motion (like the wobbling of a spinning top) so that the north pole (or rather its extension on the celestial sphere) sweeps out a circle round the pole of the ecliptic of radius $23\frac{1}{2}°$ (which is the angle between the plane of the ecliptic and the plane of the equator) in this same period of 25,800 years; the tilting is actually caused by the gravitational pull of sun and moon, which is not constant owing to the equatorial bulge of the earth. The result of this precessional movement is that the stars visible at a given locality at a particular season change considerably in the course of centuries, not only in celestial longitude (measured along the ecliptic) at the rate of 50″ a year or 1° in about 72 years (and therefore proportionately in right ascension, which is the equivalent of longitude but measured along the equator), but also in declination (the equivalent of latitude measured from the equator), because the declinations will swing to and fro 47°, i.e. $2 \times 23\frac{1}{2}°$, in 12,900 years. Hence the present appearance of the night sky in Greece (or anywhere else) is very different from what the Greeks observed; our Pole Star (α Ursae Minoris), now less than 1° from the north pole, was in Hipparchus' time (150 BC) 12° 24′ from it as he himself tells us,[6] and the vernal equinoctial point, which in his time was situated in the constellation Aries, the Ram, has now moved about 30° right out of that constellation and is in Pisces, the Fish (although modern astronomers still retain the name 'the first point of Aries' as the zero point for celestial references).

So much then for the stars. Now let us consider the sun's movement. If one observes the successive risings and settings of the sun over a period of many years, one will find that it does not rise or set at exactly the same points of the horizon at the same times each day, but that (in northern latitudes) its rising and setting points oscillate between two limits north and south of due east, and north and south of due west; on only two days of the year, in spring and autumn, does the sun rise and set due east and due west at 6 a.m. and 6 p.m. And, of course, one will discover that after $365\frac{1}{4}$ days the rising and setting sun returns to the same points of the horizon and the same times. Moreover, at its northern limit at midday the sun throws the shortest shadow in the year (i.e. its altitude above the horizon is greatest), while at the southern limit at midday it throws the longest shadow in the year (i.e. its altitude is lowest).

Furthermore, it becomes clear that as well as its daily movement from east to west the sun also has an annual movement relative to the fixed stars. This can be deduced by noting on successive nights what star or group of stars is directly overhead at, say, midnight; it will be found that they gradually change as the nights go by, until at the end of a year the same stars are found to be overhead as were there at the beginning. Now on the geocentric hypothesis (with the observer at the centre and the sun circling round him), the sun must be situated at the diametrically opposite part of the heavens to that which is actually being observed at midnight; and since this observed part changes every night, the sun's position must similarly be changing relative to the stars, until after a year it comes back to the same position again. This annual movement of the sun takes place in the opposite direction to the daily movement from east to west, so that the sun appears to be slipping back (from west to east) compared with the stars; this is why the sidereal day is 4 minutes shorter than the solar day.

Observation also shows that it is exactly the same stars each year that mark the sun's annual path; hence its circuit can be represented by a mathematical line in the heavens, defined by various groups of stars along it. This line is the ecliptic (so called because eclipses can only take place when sun and moon are both situated on it) and it forms the basic reference line in Greek astronomy; the sun's position on the ecliptic determines the season of the year – the whole circle is 360° and this is traversed by the sun in roughly 365¼ days, so that each day the sun moves a little less than 1° along the ecliptic eastwards. The ecliptic is marked by the twelve signs of the zodiac, which is defined as a narrow belt in the sky some 8° each side of the ecliptic within which the sun, moon, and planets confine their movements, the sun never deviating from the line of the ecliptic itself in the centre of the zodiac. Twelve constellations mark the zodiacal belt, the sun traversing one each month, and the following mnemonic lists these in order, starting from the vernal equinox in March:

> *The Ram, the Bull, the Heavenly Twins*
> *And next the Crab the Lion shines,*
> *The Virgin and the Scales;*
> *The Scorpion, Archer, and He-Goat,*
> *The Man who bears the Watering-Pot*
> *And the Fish with the glittering tails.*

Zodiacal *constellations* must be distinguished from zodiacal *signs*. The latter are artificially fixed segments of 30° each and by convention are unaffected by precession, so that we still speak of the sun's being in Aries at the vernal equinox, or Cancer at the summer solstice, or Capricorn at the winter solstice, although actually, owing to the effect of precession, it is no longer these constellations that mark the signs, but Pisces, Gemini, and Sagittarius. Since the sun does not always rise and set exactly due east and due west, it is obvious that the plane of the ecliptic is not the same as the plane of the equator but must be inclined at a certain angle to it; this angle is known as the obliquity of the ecliptic.

Fig. 4 shows diagrammatically the celestial sphere (which is merely a mathematical extension in space of the terrestrial sphere with equator, tropics, and the north-south axis in the same planes

2

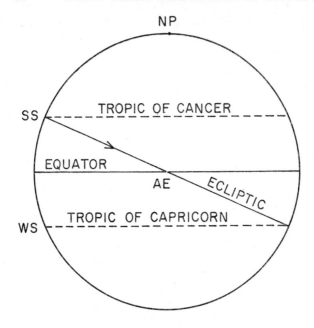

Fig. 4. Sun's apparent path on the celestial sphere

as those of the earth) with special reference to the sun. Here SS (summer solstice) and WS (winter solstice) represent the northernmost and southernmost limits of the sun's annual path ('the turnings of the sun', αἱ τροπαὶ ἡλίου) reached in June and December, when the sun is respectively in Cancer and Capricorn; AE, one point at which the ecliptic and equator intersect, represents the autumnal equinox (September), and on the diametrically opposite part of the sphere (at the back of the diagram, as it were) would be VE, the vernal equinox (March). The arrow indicates the direction of the sun's annual motion. It is important not to confuse this with its daily motion relative to a particular horizon; for each and every position of the sun on the ecliptic, it completes its daily path across the sky from east to west, but according to where it is on the ecliptic (i.e. what season of the year it is) so its daily path is different. At the equinoxes it will rise and set due east and west, and day and night will be of the same length all over the earth (hence the Greek word for 'equinox', ἰσημερία); at the summer solstice it will be highest in the sky and day will be longer than night (in the northern hemisphere), while at the winter solstice night is longer than day and the sun is lowest. In fact, the inclination of the ecliptic to the horizon is continually changing.

The ecliptic, like the equator, horizon, and meridian, is a great circle of the celestial sphere (i.e. its plane passes through the point at the centre of the sphere – in contrast to the tropics which are small circles parallel to the equator); therefore at any given moment half of the ecliptic, or six zodiacal signs, are above the horizon, and six below. If at midnight a zodiacal sign culminates (i.e. crosses the meridian, the north-south line through the observer's zenith), then we know that at midday the sun is in the sign diametrically opposite, i.e. six signs away – a fact which obviously one cannot observe directly. Furthermore, the time which the six signs, at the beginning of which the sun is situated, take to rise, marks the length of daylight, because at the end of it the sun is setting; the rising times of the zodiacal signs form an important part of ancient astronomy, and gave the impetus to the discovery of trigonometrical methods.

If one considers the relationship of the chief circles of the celestial sphere to an observer's horizon, several deductions can be made. Fig. 5 is a cross-section of the celestial sphere with the observer at the centre O, HH the horizon, EE the equator, SS the summer tropic, WW the winter tropic, P the north pole, and z the zenith point. Since latitude (or, in astronomy, declination) is defined as the angular distance of a place (or a star) north or south of the equator, all parallels of latitude being small circles of the sphere parallel to the equator,[7] it is clear that, if we know the inclination of the equator to the horizon, and hence the position of the tropics (since the arc EW = ES = the obliquity of the ecliptic), we can calculate not only the latitude but also the lengths of the longest and shortest days at that latitude, since these are determined by the ratio of the part of the tropic above the horizon to that below it at the summer and winter solstices respectively. The inclination of the equator to the horizon can be

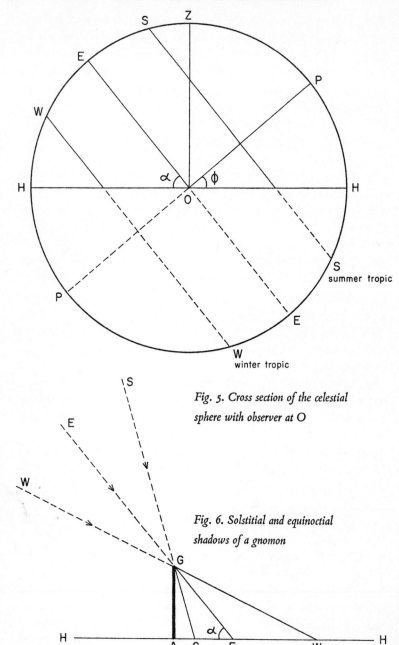

Fig. 5. Cross section of the celestial sphere with observer at O

Fig. 6. Solstitial and equinoctial shadows of a gnomon

found by measurements of the shadow of a gnomon set perpendicular to the horizontal plane, as shown in Fig. 6; here the
dotted lines represent the sun's rays striking the tip (G) of the
gnomon at the winter solstice (ww), at the equinox (EE – when the
sun is actually on the equator), and at the summer solstice (ss). If,
for example, the ratio of the gnomon to its shadow at midday on
the equinox is 4: 3[8] then

$$GA: AE = 4: 3$$
$$4AE = 3GA$$

and (in modern terms) $4 \cos \alpha = 3 \sin \alpha$

$$\cot \alpha = 0.75$$
therefore $\alpha = 53° 8'$

but $\alpha = 90° - \phi$ (see above), and hence the latitude $= 36° 52'$. In
Fig. 5, the part of the tropic above the horizon represents day and
the part below represents night; if at the summer solstice the ratio
of these parts is 5: 3,[9] then this is simply another way of saying
that the longest day is 15 hours and the shortest night 9 hours.
The method for determining the latitude from the length of the
longest day (obviously this is the same for all places on the same
parallel of latitude) may be demonstrated by Figs. 7 and 8. Fig. 7
shows a cross-section of the terrestrial globe with any point P at
north latitude ϕ on the day of the summer solstice, N being the
north pole, C the centre and R the radius of the earth; the line SC
extended in both directions represents the dividing-line between
darkness (on the left) and day (on the right), and since the time is
the longest day at this particular latitude the amount of extra
daylight may be represented by the angle OCS = the obliquity of
the ecliptic (e). From this figure we obtain the following equations:

$$OP = R \sin \theta$$
$$OC = R \cos \theta$$
$$OS = OC \tan e = R \cos \theta \tan e.$$

Now, suppose we take a different view-point and look down on
the earth from a point directly above the north pole, as in Fig. 8;
we shall now see the much greater proportion of daylight to

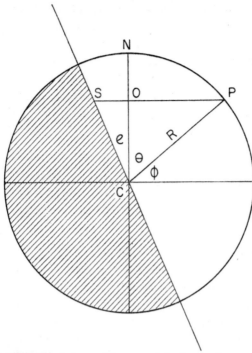

Fig. 7. Cross section of the terrestrial sphere at the summer solstice for a point P at north latitude φ

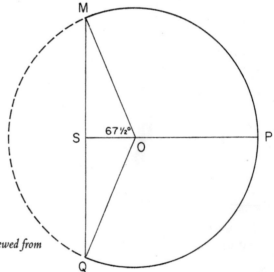

Fig. 8. Plan of the same viewed from the north celestial pole

darkness at the place P at the summer solstice, so that the dividing line will appear as MSQ, cutting off the small arc MQ representing the shortest night, and the large arc MPQ representing the longest day. If the latter is, say, 15 hours, then (since 1 hour = 15°) the obtuse angle MOQ will be 225°, and half the night arc, i.e. angle MOS, will be 67½°. Then,

$$\cos 67\tfrac{1}{2} = \frac{OS}{OP} = \frac{R \cos \theta \tan e}{R \sin \theta} \qquad \text{(from the previous figure)}$$

$$= \cot \theta \tan e$$

$$= \tan \phi \tan e \qquad \text{(since } \theta = 90° - \phi\text{),}$$

from which it follows that the latitude ϕ can readily be determined if the obliquity of the ecliptic (naturally, a constant for the whole earth) and the length of the longest day are known.

It must, of course, be emphasized that the early Greek astronomers had none of the trigonometrical knowledge assumed in the above demonstrations, nor did they express angles in degrees on the basis of the 360° division of the circle (cf. p. 157); trigonometrical methods for astronomical problems were first developed in the second century BC by Hipparchus, who was the first to compile a 'table of chords' (relating arcs of the circumference of a circle to the chords subtending them, and hence equivalent to a table of sines[10]), although the germ of such methods may be seen in the works of Aristarchus and Archimedes in the previous century. Nevertheless, it is clear that, when once the concept of the celestial sphere and the spherical earth had been established (from the end of the fifth century BC onwards), it was only a short step to the realization that the length of the longest (or shortest) day defined the position of the equator and the tropics vis-à-vis a particular horizon, and that this in turn determined the place of the observer on the earth's circumference. Certainly, Eudoxus had reached this stage of understanding (see pp. 154–55), and in the next century we find Eratosthenes using lengths of the longest day in his mapping to define the equivalents of parallels of latitude; but again it was Hipparchus who was the first to express these in degrees.[11]

There is still one easily observable solar phenomenon to be taken into consideration – namely, the inequality of the astronomical seasons, the fact that the sun takes (in round numbers) 94 days to go from the vernal equinox to the summer solstice, 92 days from there to the autumnal equinox, 89 days from there to the winter solstice, and 90 days back again to the vernal equinox. If the sun's actual course was a perfect circle with the earth as centre and if its velocity was uniform round it (both these conditions being highly desirable on *a priori* grounds), then these inequalities could not be accounted for. We shall follow the various attempts the Greek astronomers made to solve this problem, which was fundamental for the progress of astronomy.

With regard to the moon, there were two interdependent problems to be considered: (*a*) how to explain the moon's phases, and (*b*) how to frame a satisfactory theory for its motion so as to correspond with its observed appearances. The standard explanation of the moon's phases is well enough known and need not be given here; but it is worth noting that even the basic fact that the moon shines by reflected sunlight is not immediately self-evident, and it took centuries to arrive at this stage of understanding. The movements of the moon are much more irregular than those of the sun, and a mathematical theory that takes them into account is bound to be far more complicated. The moon, like the sun, moves in the general direction west to east in the zodiac, but its orbit is inclined at an angle of about 5° to the plane of the ecliptic, so that it is sometimes north of the ecliptic and sometimes south of it. Also, the two points where the moon's orbit intersects the ecliptic, called the nodes (the ascending node when the moon having been south passes to the north, and the descending node vice versa), are themselves found to move round the ecliptic, making a complete circuit in about 18⅔ years. Added to this, the so-called line of apsides joining the moon's perigee (nearest distance to the earth) and apogee (farthest distance from the earth) itself makes a complete circuit of the heavens in nearly 9 years. All these effects add up to the first inequality of the moon's motion, which in reality results from the elliptical form of the lunar orbit, the earth being at one of the foci. There is a second inequality (called the evection) and yet a third (called variation), both depending on the moon's position relative to the sun; in addition, it is known that the inclination of the lunar orbit to the plane of the ecliptic does not remain fixed but oscillates between two extremes, and there are other factors that produce significant variations in its course. In fact, the moon's motion is incredibly complicated, and its 'excessive intricacy' (to use the words of one modern historian of astronomy) is the reason why even today discrepancies are still found between the calculated and observed lunar positions.[12] We shall see that Hipparchus was able to develop a lunar theory that was sufficient to account for the first inequality, and Ptolemy (second century AD) refined on this to accommodate the second inequality and went some way towards taking into account the third.

The five planets known to the ancients may be classified in two groups – the superior planets, Mars, Jupiter and Saturn, which may be observed at different times to be at any angular distance from 0°–180° from the sun, and the inferior planets, Mercury and Venus, the angular distances of which from the sun are limited to a maximum of 28° in the case of Mercury and 47° in the case of Venus. Two further phenomena that are peculiar to the planets as seen from the earth, namely, their retrograde movements and stationary points, provide further examples of the complications inherent in trying to explain by a geocentric theory phenomena that are really only susceptible of complete explanation on a heliocentric basis. Instead of moving steadily eastwards along the zodiac relative to the fixed stars, like the sun and moon, the planets seem to slow down at irregular intervals in their courses, remain stationary for varying periods (i.e. when compared with neighbouring stars on successive nights, they show no changes in distance), and then move westwards (retrograde in the zodiac) for a time, stop again, and finally resume their eastward motion, at the same time exhibiting changes of latitude – thus in effect executing elongated loops in the sky relative to the fixed stars. We now know that the reason for these apparent motions is that the planets move in elliptical orbits with the sun at one focus, but this discovery had to await the advent of Kepler, who published it in a work that appeared in AD 1609. The development of a mathematical theory to account for the planetary movements belongs to the most sophisticated stage of Greek astronomy; Ptolemy was able to go some way towards reconciling the facts of observation with planetary theory, but the results, particularly in the case of Mars, were much less satisfactory than those achieved for the sun and moon. It is a mistaken emphasis to regard the Ptolemaic planetary theory as the crowning achievement of Greek astronomy, as is commonly done; rather is it an ingenious appendage to the primary aim, which was the establishment of a scientific treatment of solar, lunar, and stellar phenomena to facilitate the measurement of time (for which the planets are useless in naked-eye astronomy).

CHAPTER II

HOMER AND HESIOD

HISTORIANS OF THE SUBJECT are fond of saying that astronomy is the oldest of the sciences. This at first sight unexceptionable statement appears less satisfactory when we analyse the meagre evidence available for the development of astronomical knowledge in its earliest stages. Obviously, men have from time immemorial been aware of sun and moon and been grateful for the former's warmth by day and the latter's light at night. No doubt the earliest caveman was intrigued by the myriad twinklings of the stars in 'the spacious firmament on high'; probably of equal antiquity is the feeling of awe and wonder engendered by these mysterious celestial objects, leading later to the attribution of some sort of divinity to them. Yet this stage, and even the stage when individual stars and star-groups are identified and receive particular names, can hardly be characterized as 'scientific'.[13]

There would seem to be three main factors involved in the development of astronomy as a science, i.e. as an organized body of knowledge susceptible of mathematical treatment: (a) the deification of the more prominent celestial objects and the consequent study of their appearances and courses in order that rites and rituals connected with them might be properly carried out, and the supposed wishes of the deities faithfully observed; (b) the correlation of celestial with terrestrial phenomena on the assumption that the latter are caused (or at least influenced) by the former, leading again to a more detailed study of the heavens in an effort to establish a predictable pattern of events, and culminating in the predictive 'science' of astrology; and (c) the practical use of the cyclical movements of sun, moon, and stars as a means for marking the passage of time and fixing the appropriate dates of agricultural and other operations and the religious festivals connected with them – in short, the establishment of a reliable calendar. Any one of these factors to any degree, or combinations of them in any proportions, may lead to the early development of astronomical knowledge in a particular civilization; but equally, the presence of them does not necessarily presage the development of astronomy as a science.

Thus, in ancient Egypt the first factor, exemplified by the worship of the sun-god, Ra-Atum, and the sky-goddess, Nut, at Heliopolis, and the third were operative from a very early period. The heliacal rising of the bright star, Sirius, which for several centuries from about 3,000 BC nearly coincided with the beginning of the annual flooding of the Nile in the last week of June, not only served as a harbinger of this vitally important event, but also in conjunction with other stars and star-groups rising at successive intervals of 10 days (hence later known as 'decans' – see note 222) formed part of an astronomically based calendaric scheme. Yet, contrary to the belief of the Greeks themselves (*cf.* pp. 146*f.*), Egyptian astronomy never rose above the level of crude observation to develop into a science. No attempt was made to modify the decanal calendars, which were bound to become obsolete through the effect of precession (they survived for centuries as mere tomb decorations) and there is no evidence of systematic records of astronomical observations or of the application of mathematical techniques to astronomical problems (if such were recognized at all).[14]

On the other hand, all three factors may be recognized in the development of Babylonian astronomy, which certainly became scientific in character in the last four centuries BC; here it would seem that (*a*) and (*c*) were relevant from the earliest period, while (*b*) did not arise until the systematic observation of certain phenomena over many centuries had led (by *c.* 1,000 BC) to their correlation with terrestrial events, at first those affecting the country as a whole ('judicial' astrology so-called), and then much later (by *c.* 400 BC) those affecting the lives of particular individuals (horoscopic or 'genethlialogical' astrology).[15]

In the case of Greek astronomy, as we shall see, there is nothing corresponding to the solar cult of Egypt or the astral theology of

Mesopotamia; the concept of the divinity of the celestial bodies, although forming part of the official mythology and of philosophical doctrine from the fifth century BC onwards (see p. 73), never gave rise to any specific cult practices that might have had astronomical significance,[16] so that factor (a) has little relevance in this connection. It is clear that as far back as Hesiod's time the real driving force behind the progress of astronomy was (c), the practical necessity for the establishment of a reliable calendar, and it was this, coupled with the Greek passion for rational investigation along mathematical lines, that converted a mass of crude, observational material into an exact science; while (b) only became important in fostering astronomical knowledge from Hellenistic times onwards, when the tenets of astrology became firmly embedded in men's minds.[17]

We know nothing about the state of astronomical knowledge in Minoan or Mycenaean times except that, since the names of various months occur in the Linear B tablets, it seems likely that some knowledge of lunar (and presumably also solar) cycles was applied to the measurement of time. There is also evidence from Minoan and Mycenaean art that seems to indicate the worship of a sun god and a moon goddess,[18] but whether this led to any development of astronomical concepts or was accompanied by observation of astronomical phenomena for practical purposes is entirely unknown. As usual with most things Greek, one has to start from Homer, some four or five centuries after the end of the Mycenaean period, to see what can be gleaned from the *Iliad* and the *Odyssey* about the state of astronomical knowledge in the Homeric age.

Some idea of the Homeric world picture can be obtained from part of the description of Achilles' shield in the eighteenth book of the *Iliad* (483–89; cf. viii, 13–16; *Od.* i, 52–4). The earth, the shape of which is unspecified,[19] is apparently surrounded by Oceanus, the ocean river, although this is nowhere stated in so many words by Homer; it is an inference from the fact that Oceanus forms the rim of the shield (*Il.* xviii, 607–08), which is presumably round, and from the epithet ἀψόρροος, 'back-flowing' (i.e. flowing back into itself) applied to Oceanus (*Il.* xviii, 399;

Od. xx, 65). It is regarded as the source of all terrestrial waters (*Il.* xxi, 195–97), and in two passages (*Il.* xiv, 201, 246) as the origin of gods and all things – a notion that may perhaps have a non-Greek origin.[20] Above the earth is the heaven, οὐρανός, which has a stock epithet ἀστερόεις, 'starry', used even when the reference is to the daytime (e.g. *Od.* xi, 17), and is conceived of as something solid (either brazen or iron – *cf. Od.* iii, 1–2; *Il.* xvii, 425; *Od.* xv, 329; xvii, 565) which needs pillars to support it (*Od.* i, 52–4). Between earth and heaven is αἰθήρ, the upper air through which the heavenly bodies are seen on a clear night (*Il.* viii, 555–59), and ἀήρ, 'mist', 'haze', which is closer to the earth (*Il.* xiv, 287–88) – this presages the Aristotelian distinction between ἀήρ, 'air' as one of the four elements (earth, air, fire, and water – Empedocles, see pp. 52f.), and αἰθήρ as the fifth element, the substance of the highest part of the firmament (see p. 199).

Already in Homer various stars and star-groups are named. In the description of Achilles' shield (*loc. cit.*) we have the Pleiades and Hyades (two star clusters in Taurus), Orion, and Arctus, also called the Waggon (i.e. Ursa Major, the Great Bear). In *Od.* v, 271–77, we have these again plus Boötes, characterized as 'late-setting', ὀψὲ δύοντα, probably because its heliacal setting is late in the year – in 800 BC Arcturus, the brightest star in the constellation, had its heliacal setting about the latitude of Greece at the beginning of November. In *Il.* v, 5–6, Diomedes' gleaming helmet and shield are compared to the 'autumn star'; this is Sirius, which is connected with autumn because that was the time when it was most conspicuous at night, having its cosmical setting at that period round about 22 November. In *Il.* xi, 62–3, Hector, appearing here, there and everywhere on the field of battle, is compared to the same star as it appears and disappears behind clouds; it is called here 'baleful', οὔλιος, because, as we see from *Il.* xxii, 26–31 (where Achilles too is compared to this star, also known as the 'dog of Orion' from its position near this constellation), it brings fever to mortals – late summer and autumn being a notoriously unhealthy season in Mediterranean lands.

All the stars bathe in Oceanus (i.e. rise from it and return to it again) with the exception of Arctus (*Il.* v, 5–6; xviii, 489; *Od.* v,

275). This is the first reference in Greek to circumpolar stars, i.e. those which do not rise or set at a particular locality but are always visible. In Fig. 1 (p. 11) the smallest, continuous-line circle represents the limit of the circumpolar stars for that horizon, since all stars within that circle (i.e. all those with north polar distances less than the latitude of the place, or, conversely, all those with declinations greater than the co-latitude) complete their entire courses above the horizon. Now, of the seven bright stars of Ursa Major that form the well-known Plough (these were the only ones recognized at this early period as comprising the constellation Arctus – cf. Hipparchus, *Comm. in Arat.* i, 5, 6), the most southerly one (η) had in about 800 BC a declination of $+64.5°$. Greece, in an extended sense including the Aegean islands and the southern coast of the Black Sea, may be taken as lying between latitudes 33° and 43° north, so that the Plough was well within the limit of the circumpolar stars. It is described as 'alone' (οἴη) not partaking in the baths of Oceanus, because it is the only one of the star-groups mentioned, the Hyades, Pleiades, Orion, and Boötes, that is circumpolar. There is no reason to suppose that Homer was unaware that many other stars also did not bathe in Oceanus. Centuries later, the Greek astronomers included many more stars in the constellation of the Great Bear (Ptolemy lists twenty-seven) and, because the more southerly of these (those on the hind and fore feet of the animal as it was drawn) appeared just to graze the horizon without going below it in the latitudes of Greece, the circle marking off the circumpolar stars received the name of ὁ ἀρκτικὸς κύκλος, 'the arctic circle'. The use of the term is thus different from the modern sense as the limit of the north polar regions, since in the Greek sense every horizon had a different 'arctic circle' (see my *GFH*, pp. 165–66).

The sun also rises from Oceanus (*Od.* xxii, 197–98; cf. iii, 1), reaches mid-heaven (*Il.* xvi, 777; *Od.* iv, 400), and sets again into Oceanus (*Il.* viii, 485; xviii, 239–40). There is one passage (*Od.* x, 191) that speaks of the sun's going below the earth (εἶσ᾽ ὑπὸ γαῖαν), but this expression means no more than that it disappears from view below the horizon (cf. *Il.* xviii, 333, σεῦ ὕστερος εἶμ᾽ ὑπὸ γαῖαν, 'I shall go under the earth [i.e. die] later than you'), and it is quite illegitimate to assume from this a knowledge of the sun's course round and under the earth.

The day, as might be expected, is divided into three parts, morning, midday, and afternoon (*Il.* xxi, 111; *Od.* vii, 288; *cf. Il.* xvi, 777–79), and so is the night, but less specifically (*Il.* x, 251–53; *Od.* xii, 312; xiv, 483). An evening star and a star that heralds dawn are mentioned (*Il.* xxii, 317–18; xxiii, 226; *Od.* xiii, 93–4); these can be none other than the evening and morning appearances of Venus, as yet unrecognized as a single planet. Once Athene is likened to a star from which many sparks fly (*Il.* iv, 75–8) – presumably denoting a shooting star or meteor. The cardinal points of east and west are well known, defined by the sun's rising and setting (e.g. *Od.* x, 190–92). There is no mention of any analogous astronomical phenomena for north and south, but the winds Eurus, Notus, Zephyrus, and Boreas are mentioned (*Od.* v, 295–96; 331–32), and after Homer these were used to indicate direction (respectively east, south, west, and north).

In the *Odyssey*, generally believed to be somewhat later than the *Iliad*, we find mentioned (*Od.* xv, 404) the 'turnings of the sun' (τροπαὶ ἠελίοιο) and also (x, 470 – omitted in many MSS) a reference to the completion of the 'long days' (ἤματα μακρά); the first passage has been interpreted as referring to the solstices in general, and the second to the summer solstice (see Stanford *ad loc.*). In Hesiod, as we shall see, the 'turnings of the sun' definitely mean the solstices, but in the first Homeric passage (where Eumaeus, the swineherd, is telling his life-story to the disguised Odysseus) the context is the location of the island Syrie in relation to Ithaca. What is required is simply an indication of direction or distance, which would not be given by the meaning 'solstices'. The general consensus of opinion[21] is that the words are a vague expression for the far west (just as in *Od.* xii, 4, ἀντολαὶ ἠελίοιο means the extreme east, the 'risings of the sun'); but it is difficult to see why the sun should 'turn' only in the west and not the east as well. I now prefer to interpret the phrase as expressing distance rather than direction (the latter being already denoted by Ὀρτυγίης καθύπερθεν, 'above [i.e. north of] Ortygia') and as meaning the limits of the world, the extreme ends of the earth (*cf.* πείρατα γαίης, *Il.* viii, 478–79;

π. Ὠκεανοῖο, *Od.* xi, 13), where the sun disappears into Oceanus, whether west or east – to indicate at what a vast distance Syrie lies. The island, for which ancient and modern commentators have suggested localities as divergent as Sicily and Syros in the Cyclades, is, of course, a poetic fiction, quite in keeping with the poem that has given us the isles of Circe, Calypso, and Aeolus.

As to the second passage (*Od.* x, 470), there is no reason to suppose that mention of 'long days' (ἤματα μακρά) must also imply knowledge of the summer solstice – particularly as the line recurs in two other places (xix, 153; xxiv, 143) with πολλά instead of μακρά, thus giving the meaning 'When many days had come round to completion' not 'When the long days . . ..' Also in the *Odyssey* (x, 86) there is a phrase describing the country of the Laestrygonians as a place where 'the paths of day and night are close together'; this may, perhaps, be a 'muddled reference to the short summer nights of high northern latitudes' (Stanford *ad loc.*), where the sun has no sooner dipped a short way below the horizon than it rises again.[22] In both poems, references to the 'revolving year' (περιτελλόμενοι or περιπλόμενοι ἐνιαυτοί) and the recurring seasons are frequent (e.g. *Il.* ii, 551; xxiii, 833; *Od.* x, 467f.; xi, 294–95); the Seasons (ὧραι), as personified goddesses, guard the gates of heaven and Olympus (*Il.* v, 749–51), and the varying appearances of the moon are, of course, mentioned (*Od.* xiv, 161–62; 457; *cf. Il.* viii, 555–59).

So what is the total impression we gain from this survey of astronomical and related phenomena in the Homeric poems? Very much what we should expect on *a priori* grounds: at this time the obvious phenomena of the heavens were well known,[23] but simply as observed occurrences, with no hint that the reasons for them were understood or that any scientific explanation of them was even envisaged. Various stars had already been named and the more conspicuous were sufficiently familiar to be used in similes for gods and men; but the planets had not yet been differentiated. There is no indication that the heavenly bodies in general were regarded as divine, a view which later (from Plato onwards) became established doctrine. It is true that there is a god Helios and a goddess Selene, but they play very insignificant

3

roles,[24] and apparently are not thought of as causing the day-to-day appearances of sun and moon. There is only one slight hint of the idea that the stars influence human life (the fundamental tenet of astrology), and that is *Il.* xxii, 30–1 where the 'dog of Orion', Sirius, is said to 'bring' (φέρει) fever to mortals; but this is hardly more than a common figure of speech,[25] and in the same passage the star is described as an 'evil portent' (κακὸν σῆμα). On the other hand, there is abundant evidence that the correlation between the state of the heavens and the passage of time on earth was well known on a purely observational basis. The frequent references to the revolving year, mention of various constellations with descriptive epithets connected with their seasonal appearances, of the moon in different phases, of long days, of the sun's course to mark the division of the day, and so on, prove this beyond doubt, and the importance attached to such observations is imaginatively underlined by making the Season goddesses the guardians of heaven and Olympus (*Il.* v, 749–51). For the Greeks of the Homeric age onwards the desire to know how to *tell the time*, not so much the time of day or night (which came later) as the time of year, constituted the strongest impetus towards the detailed study of the heavens; it is hardly necessary to stress the importance of the seasons in a primarily agricultural civilization.

This is shown even more clearly when we consider Hesiod's poem entitled *Works and Days*. The date and authorship of the Hesiodic poems are as uncertain as those of the Homeric poems, but it is probable that the Hesiodic corpus is later than the Homeric, and certain that, although written in the same Epic dialect as the *Iliad* and *Odyssey*, the *Works and Days* is pervaded by a very different spirit. Instead of court poetry, composed for aristocratic entertainment, Hesiod gives us the earliest Greek didactic poetry; nearly half the *Works and Days* consists of a kind of versified farmer's manual, giving instructions on when to carry out various agricultural operations with a certain amount of technical advice included. Here we see that the correlation of astronomical phenomena with seasonal variations during the year has been carried a step further than in Homer (which, incidentally, confirms the view that Hesiod is later); the solstices are definitely recognized as constituting fixed reference-points in the cycle of the seasons, and observation of various stars has been carried to the extent of counting the days between their different risings and settings, giving what is in effect a calendaric scheme.

Thus in *WD* 383–87 it is recommended that the harvest should be begun when the Pleiades are rising, and ploughing when they are about to set; the reference is to their heliacal rising and cosmical setting, as is made clear by the poet's subsequent remark that they are hidden for forty days and that when they appear again 'as the year moves round' (περιπλομένου ἐνιαυτοῦ) the sickle must be sharpened.[26] Now, theoretically, the position of a star can be calculated for any epoch if due allowance is made for the effect of precession (see pp. 15*f.*) and of the proper motion (if any) of the star; hence, in theory, the times of the typical 'phases' (i.e. morning and evening risings and settings) in a star's course can be worked out for any given date and latitude. The calculations are long and laborious, even with the help of specially constructed tables,[27] and a small error in any of the constants used makes a large difference in the results; frequently the calculated times do not agree with the dates of the phases as transmitted to us from the ancient sources,[28] which themselves may give dates differing from each other by a week or more for the same rising or setting. This is not surprising in view of the difficulty of making such observations accurately; for horizon phenomena such as these, atmospheric refraction, weather conditions, and uncertainty in estimating the time and position of the first and last visibility of a star, would all militate against achieving consistently accurate results (*cf.* Ginzel, Bd. i, pp. 25–6; ii, p. 310). Add to this the fact that, with the accumulation of material by different observers in different parts of the Greek world, data which were only appropriate to one latitude were nonetheless assumed to be valid for all regions (see p. 85), and it becomes clear that comparison with the calculated dates accurate to one day (such as Ginzel provides in the table in Bd. ii, pp. 520*ff.*) is not always a fruitful procedure. In what follows I give the dates ascertainable from the ancient sources with Ginzel's in brackets (calculated by him for a latitude of 38°N. about 800 BC).

The heliacal rising of the Pleiades occurred about 5–10 May (Ginzel, 20 May), and this was regarded as the beginning of summer and the fruiting of olives and vines; previous to this, the Pleiades had been invisible from the time of their heliacal setting at the beginning of April, as Hesiod says. With advancing summer, they rose earlier and earlier before the sun until in the early morning they could first be seen setting just before sunrise, i.e. the cosmical setting, 5–11 November (Ginzel, 3 Nov.), regarded as the beginning of winter, the time for ploughing and sowing. After this they set earlier and earlier in the night, until their heliacal setting and subsequent period of invisibility. There are several other references in the *Works and Days* to this star-group, which, despite its relative dimness (the brightest star in it, η Tauri, known as Alcyone, is only of the third magnitude), is conspicuous in the night sky because of the closeness of the stars forming it, and has been used by various peoples all over the world to mark the passage of time and the seasons of the year.[29] In *WD* 571–72, the time when the snail 'flees from the Pleiades', i.e. before the heliacal rising in early May, is described as being no longer suitable for digging vineyards, and at 615–17 the Pleiades, Hyades (some 10° south-east of the Pleiades) and Orion are mentioned together as indicating by their setting that the season for ploughing has arrived. This again refers to the cosmical setting in November, with the Hyades following the Pleiades a few days later over the western horizon, and Orion beginning to set after them in the latter part of the month (Ginzel, 20 Nov.). In 618–21 the same season is referred to as marked by storms at sea 'whenever the Pleiades plunge into the misty sea to escape the mighty strength of Orion'. The heliacal rising of the latter, extending over the last two weeks in June and the first week in July (Ginzel, 29 June), is said to be the time for winnowing grain (597–98).

The two other prominent stars mentioned by Hesiod are Sirius and Arcturus. In *WD* 414–19 there is a reference to the time after the worst heat of summer, when Zeus sends the autumn rains and men's flesh feels more comfortable, 'for it is then that the star Sirius passes for a brief time during the day over the heads of men born to misery, but partakes of a greater share of the night'. The heliacal rising of Sirius took place about 20–27 July (Ginzel, 28 July), ushering in the period of greatest heat;[30] by September and October, when it was cooler, the star was visible for much of the night and had its cosmical setting about 22 November – 6 December (Ginzel, 22 Nov.). Sirius and Orion are said to be in mid-heaven at the time when 'rosy-fingered dawn gazes on Arcturus' and the grapes should be cut (*WD* 609–11); the heliacal rising of Arcturus occurred about 5–15 September (Ginzel, 17 Sept.), and at this time both Orion and Sirius would have been well up in the night sky.

Finally, we have three references to the solstices (τροπαὶ ἠελίοιο) in the poem. At 564–67 we are told, 'Whenever Zeus completes sixty wintry days after the solstice, then it is that the star Arcturus leaves the holy stream of Oceanus and first rises gleaming brightly in the evening twilight.' The achronycal rising of Arcturus took place about 24 February – 4 March (Ginzel, 24 Feb.), and sixty days back from this would make the winter solstice occur about 26 December – 3 January.[31] The same solstice is also referred to at 479–80, where a warning is given that ploughing (that is the ploughing that forms part of the same operation as sowing)[32] at the solstice will result in a poor crop – the best time being (as just mentioned) at the setting of the Pleiades. In the third passage (663–65) it is stated that fifty days after the solstice at the end of the season of heat is the time for men to go sailing; according to Kubitschek (*loc. cit.*) the summer solstice occurred on 1 July (Ginzel, Bd. ii, p. 312, note 2, says 2 July), so that fifty days afterwards would bring us to the second half of August.

We see, then, that Hesiod names almost the same stars as Homer, and evidently regards them in the same way (compare the references to Sirius and Oceanus in each author), but that he connects their phases more closely with the practical operations of life (to the extent of counting the days between various phenomena) and also relates them to the two solstices. This astronomical knowledge is based on the rough-and-ready observations, inaccurate by their very nature, unsystematically recorded and imperfectly understood, of *practical* men, farmers, sailors, and travellers, whose main concern was to have some sort of guide for the regular business of everyday life, to mark the seasons for different agricultural operations, to ensure that religious festivals were held at the appropriate times, to give warning of the months when it was unsafe to put to sea, and so on. It is not part of a system of scientific astronomy, but still belongs to the pre-scientific stage. Whether Hesiod actually wrote a work on the constellations[33] is very doubtful, but in any case it was certainly not a scientific treatise. It may have recounted some of the stories attached to the names given to certain stars and star-groups (*cf.* pp. 159f.).

CHAPTER III

THE PRE-SOCRATICS TO ANAXAGORAS

THE ACCUMULATION of such observational material and its application to the marking of the cycle of the seasons obviously continued in the centuries after Hesiod and, as we shall see, led
eventually to the establishment of an astronomically based
calendar; but before this could happen much more had to be
known about the sun's course and the relationship between its
daily rising and setting and its annual cycle among the stars. The
crude, empirical data, yielding such rule-of-thumb precepts as
'harvest when the Pleiades rise and plough when they set' needed
to be analysed and looked at in a wider cosmological setting, and
thought directed towards discovering general principles and laws
governing all astronomical phenomena. In fact, theoretical
astronomy had to be born if the study of the heavens was going
to become a science.

The first tentative steps in this direction were taken by the
Greek thinkers of Ionia in western Asia Minor in the sixth century
BC who, together with their successors and imitators in mainland
Greece and the Greek colonies of Sicily and southern Italy, are
known collectively as the Pre-Socratics. The study of Pre-
Socratic thought has proved a happy hunting ground for the
ingenious speculations of scholars both ancient and modern over
two millennia and more – a ground which has been made all the
happier because it fulfils many of the requirements for an ideal
subject for academic discussion: thus, its importance in the history
of thought is undeniable, there are *some* solid facts that are
indisputable (e.g. the names of the philosophers), much of the rest
of the evidence available to us is extremely untrustworthy
(particularly as regards the ideas of the earlier Pre-Socratics), and
the original theories relating to astronomical matters, of which we

39

catch tantalizing glimpses through the fog of second-, third-, fourth-, and fifth-hand reporting, are often fantastic and difficult of rational explanation. Add all these ingredients together, allow to simmer in the minds of credulous scholars, and some truly remarkable results may be obtained.[34]

The term 'Pre-Socratic' itself, although now hallowed by usage, is, as Guthrie points out,[35] ambiguous since it can be used in two different senses, (a) chronologically, to denote those Greek thinkers who lived before the death of Socrates in 399 BC, and (b) to denote those who were unconnected with the school of philosophy of Socrates and Plato – non-Socratics, in fact – including the sophists, some of whom were contemporary with Socrates while others were fourth-century figures. In what follows, the word is used in the second sense, largely because this was how H. Diels used it in his monumental *Die Fragmente der Vorsokratiker*, the indispensable source-book for the opinions of the Pre-Socratics (referred to hereafter simply as *DK*).[36] The sources that *DK* combs for Pre-Socratic material are extremely diverse and range in time over more than a millennium and a half – from Herodotus in the fifth century BC to Byzantine scholars of the twelfth century AD. No attempt is made at any critical evaluation of the worth of these citations, and no indication is given of the relative reliability (or lack of it) of the authors. This is not intended to be a criticism of the work; its plan in general excludes any such judgment, since it was designed to be simply a corpus of material relating to the Pre-Socratics. Nevertheless, this lack of differentiation in the sources is a factor that must be kept in mind when deciding what opinions can be attributed to individual Pre-Socratics.

It was Aristotle who initiated the practice of collecting and discussing the opinions of previous thinkers on philosophical and scientific subjects,[37] while he left to his pupils the task of compiling a comprehensive history of human knowledge. Thus Menon was allotted the field of medicine, Eudemus that of mathematics and astronomy and perhaps theology, and Theophrastus that of physics and metaphysics. It seems reasonable to assume that Aristotle and his immediate successors as mentioned above had access to most, if not all, of what had been written in Greek at least from, say, 500 BC to their own times. Their testimony, then, as to the doctrines held by individual thinkers may with some confidence be regarded as reliable since it is based on first-hand evidence – provided, of course, that this evidence, in the shape of written work, was still extant in the fourth century BC, which is demonstrably *not* true of, e.g. Thales.[38]

The situation is, however, very different for the host of epitomists, excerptors, compilers of biographical and doxographical handbooks,[39] and other popularizing commentators, from whom comes the great bulk of the passages quoted in *DK*. These writers no longer had access to the original works, which had disappeared probably centuries earlier; they took their material from the histories of Eudemus and Theophrastus (with regard to scientific δόξαι) and, what is more, not even from these *directly*, but through various intermediary 'authorities'. For the works of Aristotle's school suffered the fate of being excerpted, shortened, rearranged, and generally re-edited in the cause of popular exposition by lesser writers, who had nothing to contribute but their preconceptions of what their forerunners *ought* to have thought or said in the light of later knowledge.

The ramifications of the sources have been patiently traced in Diels' other great work, *Doxographi Graeci* (Berlin, 1879; 2nd ed. 1929), which remains the standard work in this field and some of the results of which I summarize in *CQ* 9, 1959, pp. 299*f*.[40] The essential point (which needs emphatic reiteration) is that one has to be extremely cautious in accepting, as reliable information about the Pre-Socratics, the various notions attributed to them by the later doxographers, whose accounts (garbled and frequently contradictory as they are, and nearly always coloured by later preconceptions)[41] are based on second-, third-, fourth-, fifth-, and sixth-hand evidence, which in the course of transmission has almost certainly been subject to additions, subtractions, and distortions such as to make the original ideas more or less unrecognizable. In particular, one must beware of accepting the *isolated* testimony of late writers on the earlier Pre-Socratics if this cannot be substantiated from any other source. As Kirk and Raven sapiently remark,[42] 'Thus it is legitimate to feel complete confidence in our understanding of a Pre-Socratic thinker only when the Aristotelian or Theophrastean interpretation, even if it can be accurately reconstructed, is confirmed by relevant and well-authenticated extracts from the philosopher himself' – and this is all too seldom the case.

It seems, therefore, logical to classify our information about the astronomical views of the Pre-Socratics under three main categories in descending order of reliability: (a) what we gather from the actual fragments of the writer's work where his words are evidently quoted verbatim by someone else (DK's 'B' fragments) – obviously when these are genuine they constitute the most reliable evidence for the opinions of the particular philosopher; I call this primary evidence; (b) references in the early authorities, i.e. Herodotus, Plato, Aristotle, and his immediate pupils – these may be expected to be fairly trustworthy, but already an element of uncertainty appears here, since the original notions may suffer distortion by being interpreted to conform to a specifically Platonic or Aristotelian manner of thinking; I call this secondary evidence; (c) references in all later writers (i.e. those after the death of Theophrastus, c. 287 BC) who may just possibly have had access to independent material, but are extremely unlikely to have done so, and in most cases relied on the epitomists and excerptors of the work done by the Aristotelian school – such evidence is the most likely to have suffered the maximum distortion in the process of indirect transmission, and should be used with the greatest caution; I call it tertiary evidence, although most of it is not even at third-hand but at a much greater remove from the original.

Our information about the astronomical ideas of the three earliest Pre-Socratics, Thales, Anaximander, and Anaximenes, all from sixth-century BC Miletus and constituting the so-called Milesian school of philosophy invented by Aristotle and his pupils in the second half of the fourth century BC, rests almost entirely on tertiary evidence – with the exception of two passages in Aristotle (see below). Thales was the subject of my article in CQ 9, 1959, pp. 294–309, the main conclusions of which may be recapitulated here. There is no primary evidence for him at all,

and it is certain that no written work of his was extant in the fourth
century BC. The secondary sources (Herodotus, Plato, and
Aristotle) picture him as an essentially practical man of affairs, a
sensible politician, and an astute business man, with an enquiring
mind showing a bent for natural science. It is for these qualities
that he was regularly included among the Seven Wise Men of
Greece, and his name became a household word for cleverness.[43]
Eudemus, however, used him as a peg on which to hang a notional
account of the beginnings of Greek mathematics, and it is probably
to Eudemus that we owe the picture (enlarged by later tradition
and dear to modern historians of science) of Thales as the founder
of Greek mathematics and astronomy, the transmitter of ancient
Egyptian and Babylonian science to Greece, and the first man to
subject the empirical knowledge of the Orient to the rigorous,
Euclidean type of mathematical reasoning. For this picture there
is no good evidence whatsoever, and what we know of Egyptian
and Babylonian mathematics makes it highly improbable, since
both are concerned with empirical methods for determining areas
and volumes by purely arithmetical processes, and neither shows
any trace of the existence of a corpus of generalized geometrical
knowledge with 'proofs' in the Euclidean style.

Hence the idea that Thales brought to Greece from Egypt or
Babylonia theorems proving such propositions as the equality of
the vertically opposite angles when two straight lines intersect, or
that, if two triangles have two angles and one side equal, they are
equal in every respect,[44] is wholly untenable. Similarly, he had
nothing to learn from Egyptian astronomy, which was of a crude,
empirical kind in no way superior to the Hesiodic type of astronomical knowledge described in the last chapter. The story in
Herodotus (i, 74, 2) that Thales foretold a total solar eclipse which
took place during a battle between the Lydians and the Persians
(this is generally reckoned to be the eclipse of 28 May 585 BC) cannot possibly be true as it stands, since it is out of the question that
Thales could have had the astronomical competence necessary for
such a prediction.[45] The most that can be said is that he might have
heard from Babylonian sources of an 18-year cycle (strictly, 18
years and 11 days, or 223 lunar months), in which both solar and

lunar eclipses may repeat themselves in roughly the same positions. There is some evidence that such a cycle was in use at least for lunar phenomena in early Babylonian astronomy, and Thales may somehow have connected it with a solar eclipse, so as to give rise to the story that he predicted it – in which case the fulfilment of such a 'prediction' by a total, solar eclipse was a stroke of sheer luck for him. Otherwise, there is no reason to suppose that his astronomical knowledge was very different from that of Hesiod, although doubtless by Thales' time more constellations and the risings and settings of a greater number of stars were known.

One of the main preoccupations of the Pre-Socratic thinkers was the search for an underlying, fundamental substance which could be regarded as the basic element (ἀρχή) out of which the whole universe was created – or if not a single element, a basic force or principle the workings of which regulated all the manifestations of the physical world. This entailed the abandonment of the mythological or deistic approach to the problem of creation, as recounted in Hesiod's *Theogony* 123*ff*.: 'From Chaos came forth Erebus and black Night; but of Night were born Aether and Day . . . and Earth first bare starry Heaven . . . but afterwards she lay with Heaven and bare deep-swirling Oceanus . . .' etc., etc.[46] It was undoubtedly a bold step to take and one greatly to the credit of the Pre-Socratics. Something else had to take the place of the traditional myths if a rational, coherent account of the whole natural scheme of things was to be given – hence the search for the ἀρχή. Connected with this was the attempt to look at the earth from the outside, as it were, and to consider its position in relation to the cosmos as a whole and in particular to the other celestial objects, in an endeavour to give a rational account of the whole universe and trace a logical course of development for it, independent of the stories of myth. This represents a new departure in astronomical speculation, though not in factual astronomical knowledge.

In one of the two passages referred to above, Aristotle tells us (*De Caelo* ii, 13, 295b10–12) that Anaximander regarded the earth as necessarily remaining at rest because, owing to its ὁμοιότης, 'equiformity', there is no reason why it should move in one direction rather than in another. To illustrate the contradictory nature of the sources, this passage appears as 12 A26 in *DK* and, under the same heading, *DK* includes a remark by Theon of Smyrna (a second-century AD handbook writer) to the effect that Anaximander said that the earth was a celestial object (μετέωρος) and moved round the centre of the cosmos! Whatever the exact meaning to be attached to ὁμοιότης[47] and whatever shape he imagined the earth to be (nothing certain can be inferred on this point), it is clear that Anaximander's thought was in line with the tendency to look for non-mythical explanations of cosmic phenomena, and that he was even ready to discard the notion that the earth as a whole needed anything to support it, whereas Thales (for whom the ἀρχή was water) had apparently envisaged the earth as resting on water.[48] Impressed by this intellectual daring and disposed to believe every scrap of tertiary evidence which attributes to Anaximander scientific knowledge far in advance of what was possible in his day, modern scholars have built up a vastly distorted picture of him as the first 'mathematical physicist' (no less!), endowed with superlative 'mathematical insight', and blazing a lonely trail which was not to be followed up for several centuries.[49] A refutation of such views appears in my article 'Solstices, Equinoxes, and the Presocratics', *JHS* 86, 1966, pp. 26-40, which examines the implications inherent in the assumption that Anaximander understood such concepts as the equinoxes and the use of the gnomon to measure them and tell the time of day; it is shown that such knowledge (attributed to Anaximander by one or more of the commentators of late antiquity) implies a familiarity with the concept of the celestial sphere and its main circles which is entirely anachronistic for his time, and for which there is no good evidence before the latter part of the fifth and the beginning of the fourth century BC.

The tertiary sources (e.g. *DK* 12 A18–22) also attribute to him a fantastic theory whereby he apparently conceived of the sun and moon as apertures in hoops of fire, blockage of which causes eclipses and the phases of the moon. The unsatisfactory nature of the evidence, which is garbled and contradictory and has to be interpreted with arbitrary selectivity if a coherent account is to be obtained, makes it highly doubtful whether it has any historical worth, since there is no particular reason why one scholar's favoured interpretation should be any closer to the original than another's.[50] Therefore, as this is not the place for profitless speculation, it seems best to admit that we really know nothing for certain about Anaximander's astronomical beliefs except that he regarded the earth (shape unspecified) as stationary at the centre of the universe. Whether there is any truth in Aëtius' statement (ii, 13, 7 = DK 12 A18) that Anaximander regarded the sun as the highest celestial body (i.e. the furthest from the earth), with the moon next, then the fixed stars, and then the planets (nearest the earth), we cannot tell: if this were his genuine belief, it does not say much for his astronomical competence.[51]

The other passage in Aristotle (De Caelo ii, 13, 294b13f.) states that Anaximenes, Anaxagoras, and Democritus considered that the earth kept its stationary position because of its 'breadth' or 'flatness' (πλάτος). Floating on the air beneath it, which it covered like a lid, it remained motionless with regard to the surrounding air.[52] Anaxagoras and Democritus will be discussed later; here we are concerned only with Anaximenes, the third of the Milesian philosophers, generally considered to be younger than Anaximander.[53] Evidently, the latter's unsupported earth found no favour with Anaximenes, who reverted to the idea that it must rest on something, and, since in his view the ἀρχή was air, he envisaged the earth as supported by air. We still have no clear evidence as to the shape of the earth apart from the reference to its 'breadth' or 'flatness', which would be appropriate for any shape the breadth of which is greater than its thickness. The tertiary sources attribute to him such ideas as that 'the sun is flat like a leaf' (Aëtius, DK 13 A15); that the stars are of a fiery nature and that there are also earthly bodies revolving with them (Hippolytus, DK 13 A7; Aëtius, A14); that the stars turn round (not under) the earth 'just as a felt cap turns round our head' (A7§6); and that 'the sun is hidden not by going under the earth, but by being covered by the higher parts of it' (loc. cit.; cf. Arist., Meteor. ii, 1, 354a29f. without the mention of any names). How much of this (we shall meet a good deal of it again in connection with the names of later Pre-Socratics), if any, can properly be regarded as representing Anaximenes' views is totally uncertain. A corrupt passage in Aëtius (ii, 14, 3 = DK 13 A14) speaks of the stars 'being fixed like nails in the "crystalline"' but also describes them as 'fiery leaves'.[54] Heath's interpretation of this passage, to the effect that Anaximenes must have distinguished between the fixed stars and the planets, the latter being the 'leaves' capable of being blown about in irregular fashion,[55] is very far-fetched, ignores a serious textual difficulty (cf. Guthrie, vol. i, p. 135), and goes counter to what little evidence we have as to when the distinction was made (cf. JHS 86, 1966, p. 30).

Xenophanes (c. 570–475 BC) is the first of the Pre-Socratics for whom we have any of what I call primary evidence concerning their astronomical beliefs. He wrote in verse and nearly 140 lines are quoted by various authors and listed as 'B' fragments in DK 21. However, regrettably little of this is relevant to astronomy. He said (DK 21 B32) that what men call Iris, i.e. the rainbow, was simply cloud of various colours – which is reasonable enough, since a rainbow is always seen against clouds – and that the sea was the source of all waters and winds (B30). His apparent concern with meteorological phenomena lends some support to the notions attributed to him by the tertiary sources: that the sun and stars consist of glowing clouds or else sparks, extinguished and re-kindled every day (we shall meet this idea again later); that lightning is also caused by glowing cloud; and that moisture is drawn up from the sea and forms clouds and then rain.[56] Aristotle (Met. A 5, 986b21f. = DK 21 A30) says that 'looking up into the whole sky he (Xenophanes) says that the One is God'. Elsewhere (De Caelo ii, 13, 294a21f. = DK 21 A47) Aristotle is disparaging about him, accusing him of being μικρὸν ἀγροικότερος, 'rather rustic', in his thinking, and criticizes him for suggesting that the earth is rooted in the Limitless (τὸ ἄπειρον) below it; according to Aristotle this is an easy way out of seeking the real cause. Certainly, other astronomical notions attributed to him by the tertiary sources are extremely naive, such as the suggestion that there are many different suns and moons in different parts of the earth, and that eclipses are caused by the disc's falling into an uninhabited part of the earth, which he regarded as boundless in extent (A41a; A33). On the whole, it does not seem that Xenophanes had much to contribute to the advancement of astronomy.

This is not true of Heraclitus of Ephesus (c. 540–480), who is a much more important figure in the history of philosophy in general. He believed that the basic element was fire, from which by various changes came first sea and then land (DK 22 B31). Everything is in a state of continuous change,[57] the harmony of the universe consists in a state of tension between opposites (day-night, winter-summer, war-peace, satiety-hunger), and as the world came into being from fire, so it will be consumed by fire after its due cycle. In astronomy proper Heraclitus is supposed – according to the tertiary sources *only* (there is not a trace of this in the primary or secondary sources) – to have produced a variation on a theme already attributed to Anaximander. Instead of the latter's apertures in hoops of fire, Heraclitus is said to have regarded sun, moon, and stars as actual bowls of fire (σκάφαι), the turnings and tiltings of which produce eclipses and the lunar phases; while the exhalations (ἀναθυμιάσεις), particularly from the sun's bowl, produce the differences of day and night and the warmth or coldness of the seasons (DK 22 A1 = Diog. Laert. ix, 9–11; A12, Aëtius). As with the doxographical versions of Anaximander's astronomical ideas (see above), it is very doubtful whether any of this represents even approximately what Heraclitus thought. In some respects his astronomical views seem to have been remark-ably naive; from fragments B3 and B6 we see that he regarded the sun as being about a foot wide and new each day. KR state that he was 'probably not interested in astronomy for its own sake, and seems to have been content with adaptations of popular accounts so long as his general theory of cosmological change was pre-served' (p. 203).

This, however, does not altogether do justice to Heraclitus' thought. If one examines the fragments that have any relevance for astronomy, it would seem that two things in particular struck him when he contemplated the cosmic order,[58] first the fact of its continuity, and second its periodicity. Both are emphasized in fragments B30 ('this cosmos was not made by any god or man,

but always was and is and shall be fire ever-living, kindled in
measure and extinguished in measure') and B124 (it would seem
absurd if the whole heaven and each of its parts exhibited order
and proportion in its forms and properties and cycles, but there
was none of this in the elements, 'but the cosmos in its beauty was
a heap of things piled up at random'). The periodicity, especially
of the sun, is emphasized in B94 ('the sun will not transgress its due
measures (μέτρα), otherwise the Erinyes, the ministers of Justice,
will find him out' – a somewhat unexpected concession to popular
belief) and B100, where the pre-eminence of the sun's period which
regulates the seasons (ὧραι) that produce everything is stressed (we
are reminded of the Homeric Ὧραι guarding the gates of heaven –
see p. 33). The idea of continuity is illustrated by B99 ('if it were
not for the sun, on account of the other stars it would be dark'),
which shows that he realized (as indeed did Hesiod, at least in the
case of Sirius – cf. pp. 36ff.) that the stars are always in the sky.
Similarly, the criticism he made of Hesiod's distinction between
good and bad days (*Works and Days* 765–828, a catalogue of lucky
and unlucky days), to the effect that every day has the same
nature (B106) and night and day are one (B57), may point to a
realization of the continuity of the day-and-night period (i.e.
twenty-four hours – later called νυχθήμερον) for the measurement
of time as against the natural reckoning by separate days and
nights (which we find e.g. in *Od.* v, 388; ix, 74; x, 28, *et al.*).

In B120 the reference is to the Homeric passages where the Bear
(ἄρκτος) is said to have no share in the baths of Oceanus (i.e. is
circumpolar – see pp. 30–1); Strabo (a writer on geography living
about the time of Augustus) says that Homer here uses ἡ ἄρκτος
(fem.) in the sense of what was later called ὁ ἀρκτικός (masc.), the
'arctic circle' (in the Greek sense), and that Crates (a Pergamene
scholar of the second century BC) was wrong in suggesting an
emendation of the text to read οἶος ('alone', masc.) instead of the
received οἴη (fem.). Heraclitus, says Strabo, hit off the Homeric
manner better when he speaks of ἡ ἄρκτος as forming the limits of
morning and evening, i.e. as marking off the circumpolar stars
which never rise or set from those non-circumpolar ones that do.

We come next to Parmenides of Elea (on the west coast of

4

Lucania in Italy), *c.* 512–450. Of the two main streams of Greek thought in the late-sixth and fifth centuries BC, the Ionian or eastern Greek and the Italian or western Greek, it was the latter which proved the most fertile for the development of mathematical astronomy. Parmenides wrote in hexameter verse and some 150 lines of his poem *On Nature* are extant (though not as a continuous piece). However his diction is obscure and very diverse interpretations of his meaning have been given. Basically, his position seems to be that if, as he accepted, the essential reality is One (*cf.* the Milesians' search for the primeval ἀρχή), then strict logic demands that there can be no real change anywhere and, in some sense, the whole visible world of changing appearances perceived by us is an illusion.

Parmenides' primary importance is in the history of philosophy, in that all subsequent philosophers had to take into account his logical demonstration of this seeming paradox. His astronomical ideas are not easy to discover, because the extant fragments (the primary evidence) do not give anything like a complete picture, and the doxographical tradition (the tertiary evidence) is, as usual, confused and contradictory. Nevertheless, it is doubtful whether *KR*'s judgment (p. 285 note 1), 'Whereas the astronomy of Anaximander is an appreciable part of his contribution to thought, that of Parmenides is not', is correct, since it seems to place an unjustifiably high value on Anaximander's and an unjustifiably low value on Parmenides' astronomical ideas. It appears that he conceived of the universe as a system of concentric rings or bands (στεφάναι) of fire, alternating or at any rate somehow in conjunction with other rings of darkness. Such details as there are have to be taken from a tertiary source (Aëtius, *DK* 28 A37), whose account is so compressed as to be barely intelligible.[59] Still, the στεφάναι have to be accepted as a definite part of Parmenides' thought because they are referred to in fragment B12. Here it is stated that 'the narrower ones (στεινότεραι – the adjective can hardly refer to anything other than the στεφάναι in the context) were filled with pure fire, those next to them with night, and after them hastens a share of flame; in the middle of these is the divinity (δαίμων) which guides everything.' The divinity, as we see from other contexts (B8, 14 and 30; B10, 6), was personified Justice or Necessity.

Parmenides' universe (variously called τὸ ἕν, τὸ πᾶν, τὸ ὅλον) is described in fragment B8, 42–5; it is finite, 'like the mass of a well-rounded sphere, from the middle equal in every respect'. Now this is of prime importance because it is the *first* completely authenticated assertion (there is no doubt at all that these were Parmenides' words) of the concept of *sphericity* applied to the universe as a whole – not, be it noted, to the earth itself or even to the heavens, but to the Whole. Naturally enough, the tertiary sources seize on the statement, and we are told that Parmenides was the first to assign a spherical shape to the earth and to place it at the centre of the universe (A1 and A44). This, of course, is in flagrant contradiction with his own actual words, and is yet another example of the untrustworthiness of the doxographical evidence. The idea of a spherical universe seems to have originated with the Pythagoreans, as we shall see later, and the notion of a controlling divinity placed at the centre is very similar to what we find in the cosmological system attributed to the late Pythagorean, Philolaus; it is therefore more than probable that the biographical notices that connect Parmenides with the Pythagorean school at Croton (on the west coast of the Gulf of Tarentum, south-east of Elea) are founded on fact.

The tertiary sources attribute various other astronomical ideas to Parmenides, as that he recognized that the Morning and Evening Stars were one (A1; A40a), that he considered that the sun, moon, and stars were made of compressed fire (A37–9; A41 – variations on this notion are attributed indiscriminately by the doxographers to all the earlier Pre-Socratics), and even that he first divided the earth into the five zones familiar to later geography.[60] His interest in astronomy is substantiated by fragments B10 and 11 where we find mentioned the aether, the constellations, the sun, and the 'round-eyed' (κύκλωψ) moon, the furthermost heaven, the Milky Way (γάλα οὐράνιον – the first extant mention of it in Greek), and Necessity that guides the heaven to hold the limits of the stars – a similar role is given to Necessity by Plato in the cosmological part of the myth of Er in *Rep.* x, as we shall see. Fragments

B14 and 15 both refer to the moon, which is described as a 'foreign light'[61] wandering round the earth and always looking round towards the rays of the sun; this is generally taken as evidence that Parmenides realized that the moon shines only by the light of the sun, and it certainly seems the most reasonable interpretation. Parmenides, then, is the first Greek thinker to understand the true position (at least on our present evidence). Plato (*Crat.* 409a–b) represents Anaxagoras as having put forward the view just recently (ὁ ἐκεῖνος νεωστὶ ἔλεγεν), and this is often taken to mean that he was the first to discover it; but to have said something 'recently' (νεωστί) is by no means the same as to be the first (πρῶτος) to have said it. Plato's attribution of the theory to Anaxagoras may simply have been determined by the fact that his views were much better known at Athens at this time than those of Parmenides (see below).

Parmenides' successors in the Eleatic school, Zeno (whose 'paradoxes' on space and motion are well known) and Melissus, do not seem to have been interested in astronomy – at least we have no information about their astronomical beliefs – and we may pass next to Empedocles of Acragas in Sicily (*c.* 493–433), who also wrote in verse and was strongly influenced by Parmenides' thought. Empedocles' chief claim to fame is his doctrine of the four elements, earth, air, fire, and water, and his conception of two opposing forces, which he calls Love and Strife. All sensible objects spring from the mingling of the four elements in various proportions, while the dynamic tension between Love and Strife produces perpetual cycles of change in the universe from a Parmenidean state of One Being to a state of scattered, disparate elements – Love (Φιλία or Φιλότης) tending to unite things, and Strife (Νεῖκος) to separate them. Another important concept which Empedocles seems to have initiated is that of the δίνη ('whirl', 'eddy', 'vortex') as the primal motion which apparently somehow sets in train (or is at any rate connected with) these opposing cycles: 'When Strife had reached to the lowest depth of the whirl (δίνης), and Love was in the middle of the eddy (στροφάλιγγι), under her do all these things come together so as to be one, not all at once, but congregating each from different directions at their will' (*DK* 31 B35, 3–6 – translation by *KR*, p. 346). These words are very obscure and their interpretation uncertain; but Plato's remark in the *Phaedo* (99*b*6) that 'A certain person puts a vortex (δίνην) round the earth and makes it remain stationary because of the heavens' clearly refers to Empedocles, and suggests that, in Plato's view at least, the δίνη comprised the whole of the heavens.[62] This was also Aristotle's view (see p. 199).

Empedocles' fragments provide evidence of original thinking in biology, physiology, botany, and religious matters, as well as in astronomy, although unfortunately in the latter field they give us few details of his astronomical beliefs. Like Parmenides, he envisages the One, the Whole, as a sphere 'rejoicing in its surrounding solitude' and 'equal to itself from all sides and utterly boundless' (fragments B27–B29); but, unlike Parmenides, he accommodates his sphere to the changing world of the senses by incorporating in it the four elements and two forces, which go through continuous cycles of change (B17 and B26). Here we see the Heraclitan idea of periodicity, to which I drew attention earlier, again coming into prominence, and now with special reference to cyclic and circular motion; the word κύκλος (or one of its derivatives) occurs in no fewer than seven of the sixteen fragments which possess some astronomical interest. In B17 Empedocles explains that, although the elements undergo a continual process of change, they nevertheless remain 'unmoved in their cycle' (ἀκίνητοι . . . . κατὰ κύκλον). Fragment B26 elaborates on this and B27 still further emphasizes the idea of circularity by a tautologous phrase σφαῖρος κυκλοτερής, 'a rounded sphere'. In B38 we have the aether 'binding everything together in a circle', but Empedocles does not regard this as 'limitless' (ἀπείρων) and criticizes Xenophanes for saying that it is (according to Aristotle, *De Caelo* 294a25–8). Nonetheless, in B54 (again from Aristotle) we are told that the aether 'sank with long roots down to the earth', which is very reminiscent of Xenophanes' idea that the earth is rooted in the limitless aether (see above).

The sun 'collected into a mass[63] goes round the broad heaven' (B41), and its ray 'strikes the broad circle of the moon' (B43). The latter, designated by the same phrase that Parmenides used, ἀλλότριον φῶς, is said to 'turn round the earth in a circle' (B45), to 'turn like the nave of a chariot wheel round the goal at the extremity' (B46), and to 'gaze on the holy circle opposite of its master' (B47). From all this it seems certain that Empedocles recognized that the moon circles the earth and shines by the light of the sun, and very probable that he thought of the sun as having a circular course, and of both sun and moon as circular (not necessarily spherical, but at any rate disc-like) bodies. Fragment B42[64] would seem to indicate that he knew the true explanation of solar eclipses (though not necessarily of lunar eclipses), and B48 ('the earth produces night by interposing itself to the rays of the sun') even suggests that he may have regarded the earth as a sphere, if it were not for the uncertainty produced by the tertiary evidence that credits him with a belief in two suns (on this see below).

The secondary sources add little to the astronomical information contained in the fragments themselves. Plato (*Laws* x, 889*b*) explains briefly how according to certain authorities (Empedocles undoubtedly is meant) the whole heaven and everything below it is produced by the combination of the four elements, operating not according to plan or divine dispensation, but φύσει καὶ τύχῃ 'by nature and chance' – the very opposite of Plato's own view. Aristotle (*Phys.* iv, 187a22f.), contrasting Empedocles and Anaxagoras, says that the former regarded the cosmos as produced by a cyclic process of mixture, while the latter thought of it as a single process. In another passage (*De Caelo* ii, 13, 295a16–21) Aristotle draws attention to Empedocles' use of analogy in argument,[65] and reports him as saying that the reason why the earth was not carried round by the motion (φορά) of the heavens, but stayed in its place at the centre, was the same as in the case of water in a cup which is being whirled round and yet does not spill, although the water is at times below the cup. This, of course, is an example of what we call centrifugal force, but the concept is not relevant to the stability or position of a *central* earth, as Empedocles seems naively to have imagined.[66]

The tertiary sources, as usual, are far freer and far less reliable in their attributions of astronomical ideas to Empedocles. We are told that he regarded the sun as larger than the moon, which was disc-shaped and received its light from the sun. As we have seen, this is probably correctly reported, but it is worth remarking that in the passage containing this information – Aëtius, *DK* 31 A60 – Thales, Pythagoras, and Parmenides are likewise credited with this knowledge! He is said to have believed that the heaven was made of compressed air, rendered crystalline (κρυσταλλοειδής)[67] by fire (Aëtius, A51), and that the fixed stars are closely attached to the crystalline sphere, but that the planets are loose (Aëtius, A54). This looks suspiciously like a late rationalization, as there is no other evidence that Empedocles even mentioned the planets. There are also various contradictory notices concerning alleged statements of his about the relative positions of earth, sun, and moon (*cf.* Aëtius, A50; Achilles, A55; Aëtius, A61). In two passages (Ps.-Plut., A30 and Aëtius, A56) we find an extraordinary theory that he envisaged two hemispherical suns moving round the earth, the one we see being but a reflection of the other. Very little sense can be made of this, and it is only worth mentioning because it might be the explanation of his remark in B48 to the effect that the earth produces night by interposing itself to the rays of the sun. In that case, the attribution to him of a knowledge of the sphericity of the earth is so much less likely. Whatever we may think of the tertiary evidence, it is quite clear that Empedocles was a shrewd observer of natural phenomena, and that his astronomical ideas, particularly his emphasis on circular motion, show a considerable advance on those of his predecessors.

Contemporary with Empedocles was Anaxagoras, who according to Aristotle (*Met.* A 3, 984a11 = *DK* 59 A43) was older in years than the former but 'later in his works' (τοῖς δ᾽ ἔργοις ὕστερος); his probable dates are *c.* 500–428. In dealing with the views of Anaxagoras, one has for the first time the feeling that there is solid ground under one's feet in place of the shifting sand-dunes of our interpretations of earlier thinkers.[68] This is largely because he spent a considerable part of his life in Athens, and we happen to know a good deal about Athens in the fifth century BC; it was, of course, the intellectual centre of the Greek world in its golden age and Anaxagoras had Pericles for his pupil and friend, and also taught Euripides whose plays contain several reminiscences of his teaching. Anaxagoras was instrumental in introducing the new scientific ideas to Athens, and we have it on the authority of Plato that his works were readily available. In the *Apology* Plato makes Socrates pour scorn on his accuser Meletus for suggesting that he (Socrates) corrupted the young by teaching that the sun was a stone and the moon earth: in fact these were Anaxagoras' doctrines published in books which the young could buy for a drachma at most in the market place (*Apol.* 26d). Socrates indignantly repudiates the idea that he did not believe that the sun and moon were gods 'just as other men do' (ὥσπερ οἱ ἄλλοι ἄνθρωποι), and characterizes Anaxagoras' views as 'absurd' (ἄτοπα).[69]

It is clear that his views, especially on astronomical matters, were well known in Athens,[70] and that at least one of his books was still available to Simplicius in the sixth century AD, for it is mostly from Simplicius (who was one of the more trustworthy commentators on Aristotle's works) that we get the verbatim quotations from Anaxagoras that form the extant fragments in *DK*. It is true that the same source supplies us with many of the fragments of Parmenides and Empedocles as well, but these thinkers expressed themselves in verse, which often lends itself to greater obscurity of meaning than the plain statements of prose. It is doubtful whether, even if their works had been as readily available at Athens as those of Anaxagoras (which is unlikely), the ideas of Parmenides and Empedocles would have been so easily understood by the reading public as those of Anaxagoras. From this it follows that even the information about his views retailed to us by the tertiary sources may be presumed to be more accurate in his case than for earlier thinkers – which is just as well, since the fragments themselves contain very little about his astronomical beliefs.

Anaxagoras' explanation of the cosmic order and the genesis of the universe shows recognizable affinities with earlier ideas, particularly those of Empedocles; but the latter's four elements are replaced by an infinite number of portions (μοῖραι) or seeds (σπέρματα) which in themselves contain particles of every possible substance (hence the much discussed dictum ἐν παντὶ παντὸς μοῖρα ἔνεστι, DK 59 B12). Instead of the Empedoclean forces of Love and Strife, Anaxagoras posits Mind (νοῦς) as the controlling intelligence of the whole cosmos. Mind is eternal and separate, omniscient and perfect, but not apparently divine, a sort of secular creator, as it were. This to the youthful Socrates sounded a very promising concept, and we are told in the Phaedo (97b) that he eagerly consulted the books of Anaxagoras to find out more, only to be disappointed because Anaxagoras did not in fact make use of the concept of Mind as the real cause of celestial phenomena (on this passage, see pp. 94–5). Aristotle makes a similar criticism (Met. A 4, 985a18–21; cf. DK 59 A47).

We can see the reason for these remarks by looking at Anaxagoras' own words in fragments B12 and B13. Here Mind is made responsible for originating the rotating movement of the whole cosmos (τῆς περιχωρήσιος τῆς συμπάσης νοῦς ἐκράτησεν) which initiated the process of separation that eventually resulted in the production of all the objects of the sensible world. This is Anaxagoras' version of the concept of the vortex,[71] which seems to have originated with Empedocles (see pp. 52–3) and continues to play a part in cosmogonical thinking – compare the spiral nebulae of modern astronomy. However, once the rotation has commenced, it carries on of its own accord, without apparently needing any further impetus from Mind, continuing the process of separation and indeed accelerating it; Anaxagoras actually says that the rotating movement will go on 'more and more' (ἐπὶ πλέον), and he connects it with the present rotations of the stars, sun, and moon. This is puzzling, because he can hardly have envisaged that their actual speed of rotation will increase, thereby presumably decreasing the length of the day, month, and year; perhaps he thought of the process of separation as continually extending its scope – compare the modern notion of an expanding universe. At any rate, it is clear that the idea of circular motion, which was emphasized in Empedocles' astronomical thought, is also prominent in Anaxagoras'. There is only one other piece of astronomical information contained in the fragments themselves, namely that the sun gives the moon its light (B18); also B19 shows that he realized that the rainbow is reflected sunlight.

The secondary sources add a certain amount. Plato confirms that Anaxagoras knew that the moon shines by reflected light and that it circles the earth (*Crat.* 409a–b = DK 59 A76). From the same passage it appears that he may also have invented the term ἕνη καὶ νέα for the last day of the month (literally, 'the old and the new' – *cf.* Aristophanes, *Clouds* 1134). Aristotle says that Anaxagoras regarded the Milky Way as starlight which is perceptible to us when the sun is carried below the earth and the latter intercepts the sun's rays (*Meteor.* i, 8, 345a25f. = A80), that he explained comets as the conjunction of planets (at last we have definite evidence that these were at least known) when they seem to touch each other (*ibid.* i, 6, 342b28 = A81), lightning as the kindling of the fiery aether brought down to near the earth, thunder as the noise of the quenching of this fire (A84), and earthquakes as the result of pockets of aether being entrapped in the hollows of the earth (A89). It seems certain that Anaxagoras envisaged the earth as being at the centre of the cosmos, brought there by the action of the primal motion (A88). It is even possible that he regarded it as a sphere, since Aristotle (*Meteor.* ii, 7, 365a23), quoting him (in connection with earthquakes) as using the phrases 'the upper and lower parts' of the earth, adds an explanation that the upper part 'of the whole sphere' (τῆς ὅλης σφαίρας) is the part on which we dwell. Unfortunately, we cannot be certain here whether this explanation is Anaxagoras' or Aristotle's, and in view of the tertiary evidence below it would seem evident that it was Aristotle's.

The tertiary sources add a good deal more, and I give a selection of attributions which are more likely to be correct than not. The sun is a fiery mass of ore bigger than the Peloponnese (Diog. Laert. ii, 8) – for this heresy Anaxagoras is said to have been banished from Athens.[72] The earth is flat (πλατεῖα) and not hollow and remains in position supported by air (Hippol., *Ref.* i, 8, 3 = DK 59 A42). Once Simplicius (*in De Caelo*, p. 520, 28) attributes a 'drum-shaped', τυμπανοειδής, earth to Anaxagoras, as well as to Anaximenes and Democritus, but this idea is usually connected with Leucippus, as we shall see. Sun, moon, and all the stars are fiery stones included in or embraced by (συμπεριληφθέντας) the revolution of the aether (Hippol., *op. cit.* i, 8, 6); beneath the stars revolve, with the sun and moon, other bodies unseen by us (*ibid.*). The heat of the stars is not perceptible because of their great distance from the earth (*ibid.* 7). The moon is closer to us than the sun (*ibid.*). The revolution of the stars takes them under the earth (*ibid.* 8). The moon is eclipsed by the interposition of the earth and sometimes of the other bodies between it and the sun (the other bodies were evidently postulated to account for the greater frequency of lunar than solar eclipses), and the sun is eclipsed at new moon when the moon interposes itself between earth and sun (*ibid.* 9). Both sun and moon make 'turnings' (τροπαί), being driven off their courses by air, and the moon often 'turns' because of its inability to overcome the cold (*ibid.*)! Anaxagoras was the first to give the true explanation of the moon's phases (*ibid.* 10), but no details are given here or elsewhere beyond this bare statement. He explained hail as pieces of compressed cloud expelled down to the earth and made round (στρογγυλοῦται) in their descent, and sensibly suggested that the Nile's flooding was caused by the melting of snow in Ethiopia in the summer (Aëtius, A91). Originally, Anaxagoras thought, the stars revolved as round a dome (θολοειδῶς) and the ever-visible pole (i.e. the north pole) was directly over the earth (κατὰ κορυφήν), but later the cosmos became tilted towards the south (Diog. Laert. ii, 9). This seems a recognizable attempt to account for the facts that the plane of the ecliptic is inclined to the plane of the equator (days and nights *not* being equal all over the world), and that in Mediterranean regions the point in the sky about which all the stars are seen to revolve (i.e. the celestial north pole) is not at the observer's zenith.

From all this evidence it would appear that at least some astronomical thought in the latter part of the fifth century BC was beginning to move away from the speculative theorizing of the earlier thinkers towards a more empirical attitude which was prepared to take into account the facts of actual observation. By this I do not mean to suggest that the results of observation were actually ignored by the earlier Pre-Socratics; but it is difficult to resist the feeling that their ideas on the physical universe *olent lucerna*, 'they smell of the lamp' – they are the dream children of the speculative thinker in his study intoxicated by the novelty and daring of the new intellectual atmosphere, and intent on applying the new methods of thought (based on abstract reasoning and the free expression of imaginative ideas) in the widest possible field. This was the real service done by the Pre-Socratics and constitutes their undeniable importance in the history of thought: by abandoning mythological traditions and subjecting external phenomena to a process of logical reasoning, untrammelled by religious dogma, and even by investigating the actual processes of thought, they opened up a whole new field of knowledge which is virtually inexhaustible. They were *not*, however, primarily scientists, much less astronomers, and observation of actual celestial phenomena seems to have played a relatively minor role in their thinking.[73] Yet, observation is fundamental to any scientific astronomy, because it is only on the results of observation that one can base a mathematical theory that will describe celestial phenomena in the *quantitative* terms necessary to attain what was said earlier to be the main goal of ancient astronomy, the measurement of time.

If one studies the various astronomical schemes put forward (or alleged to have been put forward) by the Pre-Socratics, one striking characteristic stands out as common to all of them, namely that they are all *qualitative* in conception – the notions are expressed in descriptive language apparently unconnected with terms of measurement, and not based (or at any rate very tenuously based) on any observational parameters. For the development of mathematical astronomy what was now required was much closer attention to observational material, in the shape of records of lunar and solar cycles, of the risings and settings of stars, and of the solstices. In fact this is the type of material which we find in the Hesiodic poems, and it must be envisaged as steadily accumulating all through the sixth and fifth centuries BC, largely independent of the speculative thinkers. This quantity of empirical knowledge needed to be fitted into a general picture of the universe as a whole, a celestial model, in which the earth would occupy a special position as the standpoint of the observer, and the sun, moon, stars, and planets would follow their individual courses inside the general framework. Such a framework would provide the essential system of reference to which all observational material could be related, and which would, in conjunction with more accurate methods of physical measurement, facilitate the mathematical treatment of empirical data, and so lead to the further refining of astronomical theory. As everyone knows, the general framework or model that was eventually adopted was the celestial sphere with the spherical earth set immovably at the centre; this was to prove a remarkably long-lived concept (entirely Greek in origin), and is still in use today, providing the essential basis for such things as star-maps, celestial globes, and planetaria, which by their nature require the observer to occupy a central position – although, of course, we now know that the earth is not the centre of the universe, but merely a small planet revolving round rather a dim star in one of the lesser galaxies.

CHAPTER IV

THE PYTHAGOREANS AND LATER PRE-SOCRATICS

As we have already seen, the concepts of periodicity, sphericity, and circular motion as applied to the universe as a whole are already present in the work of Heraclitus, Parmenides, Empedocles, and Anaxagoras; but astronomical thought is still in the pre-scientific stage, and the idea of the celestial sphere is still far from being worked out in detail. The nearest approach in the Pre-Socratic period to an astronomical scheme which might possibly have taken the place of the concept of the celestial sphere is in the system attributed to the later Pythagoreans of the last half of the fifth century. Now it has been said (Guthrie, vol. i, p. 146) that 'The history of Pythagoreanism is perhaps the most controversial subject in all Greek philosophy', and 'No one can claim even to have plumbed what a modern scholar has despondently called "the bottomless pit" of research on the Pythagoreans'. Here we are concerned only with the role of the Pythagoreans in the history of astronomy; but even to elucidate this some attention must be paid to the difficulties peculiar to any discussion of Pythagorean beliefs.

Put very briefly, the main problem is to try to distinguish the teachings and discoveries of Pythagoras himself from those of his successors in the school which he founded, and which continued as a recognized body of doctrines some two hundred years after his death – and to differentiate between these successors. The problem is complicated by several factors: (a) Pythagoras himself, who founded the Pythagorean brotherhood in the latter part of the sixth century BC at Croton in south Italy, left no writings; (b) it was a religious and mystical association or sect, as well as a philosophical school, which had strict rules of secrecy concerning its teachings; (c) there was a strong tendency for later writers to attribute as many discoveries as possible to the Master, who rapidly became an almost superhuman figure – unhistorical legends about him were already circulating in Aristotle's time; (d) the original brotherhood, which had by this time set up branches in many towns of Magna Graecia, was forcibly disbanded in the middle of the fifth century, and its members dispersed throughout Greece carrying their teachings with them; and finally (e) the sources which purport to tell us most about the movement, namely the so-called Lives of Pythagoras by Porphyry and Iamblichus (Neo-Platonists of the third and fourth centuries AD), at best go back to fourth-century BC accounts which were already uncritical in character (as just mentioned), and at worst are patent amalgams of Pythagorean, Orphic, and Neo-Platonic doctrines. In fact, as in the case of Thales, the tertiary sources are utterly unreliable for Pythagoras himself.

Of the secondary sources, Herodotus mentions the Pythagoreans once (ii, 81) and Pythagoras once (iv, 95), both times in connection with the doctrine of transmigration and the immortality of the soul. Plato, who was undoubtedly greatly influenced by Pythagorean teaching, and not least in his astronomical ideas (as we shall see later), yet names Pythagoras only once (Rep. x, 600b, where he contrasts him favourably with Homer for setting a good example for his followers to emulate), and the Pythagoreans once (Rep. vii, 530d, where Socrates draws attention to the importance attached by the Pythagoreans to the sister sciences of astronomy and harmonics–see pp. 108f.). Aristotle, in the extant works, makes only two incidental references to Pythagoras, one of which occurs in a passage where the text is disputed (Met. A 5, 986a30), and neither of which gives any information about him. (The other passage is Rhet. 1398b14, where the name occurs in a sentence illustrative of inductive reasoning.) On the other hand, Aristotle frequently mentions the Pythagoreans, whom he sometimes designates as οἱ καλούμενοι Πυθαγόρειοι, 'those called Pythagoreans', and he is, in fact, our most reliable authority for their scientific teachings. He wrote a work Περὶ τῶν Πυθαγορείων which has, unfortunately, not come down to us.

From the state of the evidence, then, it would appear to be a hopeless task at this distance of time to try to apportion various doctrines to individual Pythagoreans, when even Aristotle was not in a position to do so. He knew that different opinions were current among these philosophers, but he was apparently unable to connect them with particular names, and has to be content with saying 'some of the Pythagoreans think this . . . while others think that . . .' – whereas in dealing with the other Pre-Socratics he nearly always attributes specific doctrines to individual thinkers by name. Attempts by modern scholars to differentiate between what they suppose to be the views of Pythagoras himself and those of his pupils cannot therefore rest on any firm basis and should be discounted. In particular, Heath's ideas on the state of astronomical knowledge in Pythagoras' time (sixth century BC) are far too sanguine. He says (*Arist.*, p. 51),

> It appears probable, therefore, that the theory of Pythagoras himself was that the universe, the earth, and the other heavenly bodies are spherical in shape, that the earth is at rest in the centre, that the sphere of the fixed stars has a daily rotation from east to west about an axis passing through the centre of the earth, and that the planets have an independent movement of their own in a sense opposite to that of the daily rotation, i.e. from west to east.

As we have seen, there is no good evidence for such advanced knowledge in this early period, and historically this picture is completely anachronistic (on this see further my articles on Thales and Anaximander). The best that we can do is to follow Aristotle's example, and examine the astronomical knowledge which there is reasonably good evidence that the Pythagorean school in general possessed, bearing in mind that most of it must be referred to the latter part of the fifth century BC.

The great novelty that the Pythagoreans introduced into Greek philosophical thought as a whole was their insistence on the importance of number. Number was for them the ἀρχή that the Ionian philosophers sought. Numbers were the basic stuff (ὕλη) out of which the whole universe developed – numbers, in some way which Aristotle was never able to fathom, and which he evidently and with good justification regards as extremely perverse (*cf. Met.* A 8, 989*b*29*ff.* and N 6, 1092*b*26*ff.* = DK 58 B22 and B27), conceived of as actual physical entities occupying spatial extension, not intellectual abstractions; and numerical relationships constituted the governing principles of the entire cosmos. This is not the place to go into the details or speculate on the origin of this remarkable doctrine. Perhaps the discoveries (traditionally ascribed to Pythagoras himself) of the chief musical intervals, the octave, fourth, and fifth, expressed as the numerical ratios 2:1, 4:3, 3:2, of the fact that the first four integers add up to 10 (called by the Pythagoreans the 'tetractys' and regarded as a sacred symbol), and that the diagonal of a square is incommensurable with its side (the famous 'theorem of Pythagoras'[74]), all contributed to their belief that numbers formed the fundamental elements of the universe – but this belongs properly to the history of mathematics.

In astronomy the Pythagoreans also introduced a startling innovation. This was to displace the earth from the central position in the cosmos which it seems to have occupied in the astronomical ideas of most of the Pre-Socratics, and to regard it as another celestial body moving in a circle, like the sun, moon, and stars, round a central fire which provided the motive power for the whole universe and was variously called the watch or the tower or the throne of Zeus, or the hearth of the universe. Aristotle tells us (*De Caelo* ii, 13, 293*a*23 = DK 58 B37; *cf.* also Simplicius *ad loc.*) that, as well as the earth, the Pythagorean scheme postulated a counter-earth (ἀντίχθων) which also moved round the central fire, closer to it than the earth, but always invisible to us because we live on the hemisphere facing away from the counter-earth.[75] Aristotle's account does not mention the planets by name; we are merely told that the Pythagoreans regarded the earth as 'one of the stars' (ἕν τῶν ἄστρων), but ἄστρον is a general word (as indeed is ἀστήρ) which can be applied indifferently to the fixed stars, the planets, the sun, and the moon. A tertiary source (Aëtius, *DK* 44 A16 and 17) fills in some details and specifically attributes the scheme to Philolaus, a Pythagorean of the latter half of the fifth century BC.

Under Philolaus (*DK* 44) *DK* lists nineteen passages (B1–B19) as definite fragments of a work Περὶ φύσεως.[76] Unfortunately, the genuineness of these is disputed (*KR*, p. 311, will have none of them; Guthrie, vol. i, p. 331, is far less convinced they are false), but anyway only one fragment (B7) is apparently connected with this astronomical scheme, and all *that* says is that in the middle of the sphere is what is named the hearth (ἐν τῷ μέσῳ τᾶς σφαίρας ἑστία καλεῖται). According to Aëtius (*DK* 44 A16 and 17), in the middle was the central fire, next came the counter-earth, then the earth, then the moon, then the sun, then the five planets, and finally the outer sphere of the universe which carried the fixed stars. It is noteworthy that the order of the individual planets is not specified, nor are they actually named; in fact, the Greek names for Saturn, Jupiter, and Mars, i.e. the stars of Cronus, Zeus, and Ares respectively, appear only in the *Epinomis* (987c) of the extant texts before Aristotle,[77] although Plato in the *Timaeus* (as we shall see) mentions the Morning Star (Venus) and Mercury, and knows that the planets are five in number (38c–d). If we assume that the order of the planets was that generally accepted in later Greek astronomy (which is the order of their sidereal periods and of their distances from the sun), then the complete Philolaic scheme will have been (starting from the outside, τὸ περιέχον) fixed stars, Saturn, Jupiter, Mars, Venus, Mercury, sun, moon, earth, counter-earth, central fire – making ten bodies in all (counting the fixed star sphere as one) moving in circles round the fiery centre.

The postulate of the counter-earth is a puzzling feature of the system. In one place (*Met.* A 5, 986a3f. = *DK* 58 B4) Aristotle suggests that the only reason for it is to bring the total number of celestial orbits up to ten, which was the sacred number of the Pythagoreans, the visible orbits being only nine, i.e. fixed stars, five planets, sun, moon, and earth. In another passage (*De Caelo* ii, 13, 293b21f.) he mentions the view held by 'some' that there were other bodies, invisible to us, also encircling the middle (φέρεσθαι περὶ τὸ μέσον), which were supposed to explain why lunar eclipses are more frequent than solar – an idea, as we have seen, already attributed to Anaxagoras. It has been suggested that the counter-earth might have performed the same function in the Philolaic scheme. This, however, is impossible, since the orbit of the moon is outside the earth's, while that of the counter-earth is inside it (i.e. nearest the central fire). The other bodies mentioned by Aristotle were presumably envisaged as having their orbits between those of the moon and the earth if they were to produce more frequent lunar eclipses, but the counter-earth itself could not have had this effect.[78] Simplicius, in his commentary on the above passage of the De Caelo, says that the Pythagoreans called the moon ἀντίχθων and also a 'heavenly earth' (αἰθερίαν γῆν). There is nothing intrinsically improbable here (pace Guthrie, vol. i, p. 291); both moon and counter-earth encircle the central fire, one on either side of the earth, so that either could be described in some sense as ἀντί, 'over against', 'counter to' the earth.[79] In the same passage Simplicius informs us that 'those who are familiar with the more genuine doctrine call fire in the centre the creative force which from the centre gives life to the whole earth and warms again that part of it which has grown cold'.[80] The only difference between this and Aristotle's version is that the latter speaks of the central fire without mentioning its 'creative force' and its effect on the earth – presumably, knowledge of this constitutes the 'more genuine' doctrine. Perhaps a more cogent reason for the postulate of the counter-earth was to account for the fact that we never see the central fire, and to exclude the possibility that some daring explorer might go round to the other side of the earth and wonder why he still could not see the fiery centre, which, as consisting of divine fire that provided the generative force for the whole cosmos, might be expected to be a spectacular sight. Assume the existence of a body like the counter-earth which moves with our earth at the same speed in an orbit between us and the central fire,[81] and at least you have a specious reason for our never seeing it.

Aristotle says specifically that according to the Pythagoreans 'the earth being one of the stars and carried in a circle round the centre makes (ποιεῖν) night and day' (De Caelo 293a22–3), and this is duly confirmed by Simplicius who adds that night and day depend on the position of the earth relative to the sun. Exactly how is not explained, but the assumption that the earth was supposed to take 24 hours to circle the centre, the moon 29½ days, and the sun a year would suffice to account very roughly for some of the main phenomena of the heavens as seen from the earth. Day would be produced when the inhabited part was facing the sun on the same side of the central fire, and night when it had moved round 180° to the other side of the centre. This is presumably what Simplicius means when he goes on to say (p. 512, 16–17) that night results from the earth's coming into the cone of its shadow: since we live only on one hemisphere, night is produced when the other hemisphere faces towards the sun and casts its shadow over our side. The phases of the moon would be approximately accounted for: in Fig. 9, position (a) would be near new moon and (b) and (c) near full moon, but in between the latter positions (when the moon was on the line joining sun, earth, counter-earth, and central fire) it would seem that for a terrestrial observer it must disappear briefly and then reappear. Again every month there would have to be a solar eclipse, and days and nights would have to be of equal length always all over the world.

Aëtius says that Philolaus regarded the sun as glassy (ὑαλοειδής), 'receiving the reflection of the fire in the cosmos, and filtering both light and warmth through to us, so that in a certain sense there are twin suns, the fiery one in the heaven (ἐν τῷ οὐρανῷ) and the fiery one by reflection from it' (DK 44 A19). Unfortunately, it is not clear which fire is meant here, for Philolaus seems to have envisaged two sources of fire, the outermost surrounding fiery sphere (this originally Heraclitan idea was, according to the doxographers, common to several later Pre-Socratics) and the central fire (DK 44 A16); anyway, it is difficult to see how a reflected sun really helps matters.

Another minor difficulty (of which Heath makes much, Arist., pp. 101f.) is that, according to Aristotle and the tertiary sources, in the Pythagorean scheme there were ten bodies moving in ten orbits round the central fire; but if the daily rotation of the heavens is to be accounted for by the earth's moving round the central fire, there is no need to postulate any motion of the fixed star sphere, so that there would have been only nine circular motions,

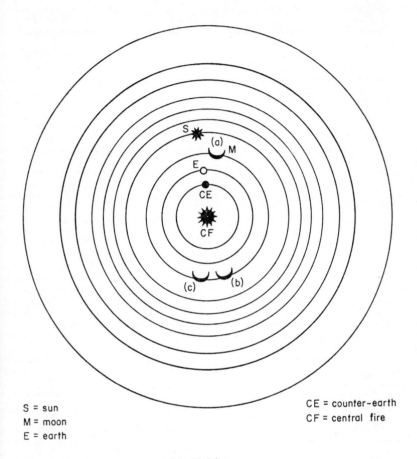

S = sun
M = moon
E = earth

CE = counter-earth
CF = central fire

Fig. 9. Schematic representation of the Philolaic system

contrary to what all our sources affirm. To resolve this discrepancy it has been suggested that the sphere of the fixed stars was, indeed, endowed with a very slow revolution, and that this must have been intended to represent the phenomenon of the precession of the equinoxes (see pp. 15*f.*). However, to suppose that astronomical theory or observational technique had reached such a level in Philolaus' time that the effects of precession (about 50″ of arc a year for stars on the ecliptic) would be noticed, is quite out of the question, and it is now certain that it was Hipparchus in the second century BC who made this discovery. The inventors of the counter-earth and the central fire (neither of which has any observational basis) could well have assumed a very slow rotation of the outer sphere, which could anyway not be disproved,[82] and which would bring the total number of revolutions up to ten and would fit in well with their preconceived notions on the construction of the universe.

The whole scheme is a good example of the type of presumptive theorizing that characterizes much of the astronomical thinking of the Pre-Socratics; it is very much a product of the study and bears little relation to the facts of actual observation (e.g. it takes no account of the latitudinal and longitudinal variations of the planetary bodies). Van der Waerden's belief that the scheme, which he describes as 'eine so raffiniert ausgedachten astronomische System', must belong to a stage in astronomical thought *after* the *Timaeus*, i.e. not before *c.* 360 BC, is wholly untenable.[83] Rather is it just the sort of scheme with which Plato was familiar from his knowledge of Pythagorean doctrines, and which he unsuccessfully tried to reconcile with the slightly later concept of the celestial sphere and the spherical earth at the centre (see below, Chapter V, especially p. 149).

The fatal objection to it (as so far described in the sources) is that it apparently envisages the orbits of the earth and all the heavenly bodies round the central fire as being in the same plane, whereas, of course, to correspond with observed phenomena, the plane of the orbits of sun, moon, and planets should be inclined to that of the earth. There is only one mention of inclined orbits in our sources and that is by Aëtius (*DK* 44 A21) where he states, 'Some say that the earth is at rest; but Philolaus the Pythagorean says that it is carried in a circle round the fire on a slanting circle in a similar fashion to the sun and moon' (κύκλῳ περιφέρεσθαι περὶ τὸ πῦρ κατὰ κύκλον λοξὸν ὁμοιοτρόπως ἡλίῳ καὶ σελήνῃ). As it stands, and if the whole sentence is to be attributed to Philolaus, this can hardly mean anything other than that *all three* bodies, earth, sun, and moon, move in an inclined orbit. Yet this would not produce the required results. It is possible that the last four words are not Philolaus', but added by Aëtius or his source, and this apparently is how Heath takes it when he says, 'The earth revolves round the central fire in the same sense as the sun and moon (that is from west to east), but its orbit is obliquely inclined; that is to say, the earth moves in the plane of the *equator*, the sun and the moon in the plane of the zodiac circle. It would no doubt be in this way that Philolaus would explain the seasons.'[84] Clearly, this entails reading a great deal more into the evidence than is in fact there. Ὁ λοξὸς κύκλος was in later Greek astronomy a normal expression for the ecliptic; it seems more than probable that Aëtius (or his source), knowing this, added it to the astronomical knowledge attributed to Philolaus simply to make the latter's views sound more plausible.[85] It is noteworthy that in the longer passage of Aëtius that describes the complete scheme (*DK* 44 A16), and in Aristotle's and Simplicius' account, there is no mention at all of inclined orbits, and it would seem very questionable to accept the unsupported statement of a tertiary source such as Aëtius in a case like this.

Very little astronomical sense is apparent in another feature of the Pythagorean scheme, the famous 'harmony of the spheres'. According to Aristotle (*De Caelo* ii, 9, 290b12f. = *DK* 58 B35), an absurd and extravagant opinion (his own words) was held by the Pythagoreans, to the effect that, with so many huge celestial bodies whirling at such great speeds round the centre, it was impossible that no noise should be generated by their motions, but each body must produce a different tone according to its distance from the centre, so that the whole system created a 'harmony'. An ingenious explanation was given of the awkward fact that no one ever hears this harmony, namely that everyone, from the moment of birth, has this sound as a constant background and therefore does not consciously hear it, since there is no absolute silence to contrast with it. This poetic fancy (presumably suggested by the discovery of the ratios governing the chief musical intervals – see above) was taken up by Plato in the myth of Er in *Republic* x (see p. 111), and further elaborated by later writers who invented all sorts of musical schemes supposed to represent the proportionate distances of the heavenly bodies – Heath (*Arist.*, pp. 105f.) treats of these at some length. Nevertheless it has no significance for mathematical astronomy, however much it gripped the imagination of later ages, except in so far as to demonstrate once more how prone the Pythagoreans were to subordinate the facts of natural phenomena to their philosophical and mystical predilections.

Despite the many drawbacks of the Pythagorean scheme, it also displays several features which permanently influenced Greek astronomical thought. It emphasized the circular motions of the heavenly bodies round a common central point; the concept was already prominent in Empedocles' ideas, but the Philolaic system went a step further by actually postulating an imaginary centre of rotation, and this was destined to be a key concept in the later development of the epicyclic and eccentric theories. It differentiated between the planets and the other celestial objects, and there is no good evidence that this distinction (traditionally, but improbably, credited to Pythagoras himself) was made by any of the earlier Pre-Socratics except Anaxagoras. Above all, it accustomed men to thinking in terms of the sphere as what one might call the typical astronomical shape, not only for the universe as a whole (this had been generally accepted since Parmenides), but more particularly as the shape of the earth itself. After the fifth century BC no Greek writer of any repute (with the exception of the atomists, Leucippus and Democritus, who held somewhat reactionary views on astronomy – see below) conceived of the earth as anything other than a globe. In Socrates' youth the shape of the earth was still a debatable matter (*cf.* p. 94), but the great authority of Plato, who certainly believed in a spherical earth (see p. 98) and who did much to make Pythagorean astronomical notions respectable, was sufficient to put the matter beyond doubt. Aristotle accepts the sphericity beyond any question (*De Caelo* ii, 14, 297a8ff.) and dismisses the proponents of a flat or drum-shaped earth in a few sentences (*ibid.* 294a).

That the Pythagoreans *did* regard the earth as a sphere is certain. In a curious chapter of the *De Caelo* where Aristotle is discussing which is 'upper' and 'lower' and which is 'left' and 'right' in the universe (ii, 2), he takes the (to us) unnatural view that the invisible (i.e. south) pole is the upper one, and those living there are in the upper hemisphere and on the right, while we (i.e. the inhabitants of the northern hemisphere) are in the lower hemisphere and on the left, 'contrary to what the Pythagoreans say; for they make us the upper ones in the right-hand part, and those at the south pole the lower ones in the left-hand part' (285b25–7). However odd the argumentation appears to us,[86] it is at least clear that the Pythagoreans considered the earth as spherical. This is confirmed by one of the reasons that Aristotle reports for the Pythagorean view that the earth and the counter-earth move round the centre: he says (293b25–30) that they considered that it made no difference to the observed phenomena, since, even on the assumption that the earth (not the central fire) was at the centre of the universe, we ourselves anyway lived half a diameter of the earth away from its centre. Thus the sphericity is again assumed, although the truth of the Pythagorean argument entails ignoring completely the different effects of parallax in the two cases, a procedure which Heath justifiably describes as 'a somewhat extreme case of making the phenomena fit a preconceived hypothesis' (*Arist.*, p. 100).

Another very influential tenet of Pythagorean thought was their insistence on the divine nature of the celestial bodies. This also received the sanction of Plato and Aristotle, became orthodox doctrine which was accepted even by the mathematical astronomers (see the introduction to Ptolemy's *Almagest*, ed. Heiberg, pp. 6, 23; 7, 20*ff.*), and provided the fundamental basis for astrology. There is little doubt that the idea of the divinity of the planets, at any rate, came originally from Babylonia, where an astral religion is attested as early as the second millennium BC.[87]

According to the tertiary sources, two individual Pythagoreans from Syracuse, Hicetas and Ecphantus (*DK* 50 and 51), introduced a modification of the Philolaic system by postulating a central earth rotating on its own axis from west to east, thus accounting for the daily movement of the heavens. Cicero (*Acad. Prior.* ii, 39, 123), ostensibly quoting Theophrastus, would have us believe that Hicetas regarded the earth as the only moving body in a universe where the sun, moon, and stars were all stationary. This, if true, would demonstrate an unbelievably imperfect knowledge

of astronomy on Hicetas' part, since it would argue that he completely ignored the proper motions of the planetary bodies in the
zodiac. However the tertiary evidence is very meagre, and even
the actual existence of these two Pythagoreans has been doubted
(*cf.* Guthrie, vol. i, p. 323).

As to other astronomical knowledge attributed to the Pythagoreans, Aëtius reports that Philolaus regarded the moon as like the
earth, but with animals and plants fifteen times the size of terrestrial ones and a day fifteen times as long (*DK* 44 A20); this is
obviously an inference from the length of a month. Diogenes
Laertius says that they knew that the moon was illuminated by the
sun (*DK* 58 B1*a*), which is probable enough as it was certainly
known by Anaxagoras; on the other hand, according to Aëtius
(*DK* 58 B36), some of the later Pythagoreans still thought of the
moon's waxing and waning as caused by its own flame – which
indicates little awareness of the true nature of lunar phenomena.
From the concept of the spherical universe and the spherical
earth, it would seem but a small step to the acceptance of the idea
that all the celestial bodies are spherical; but there is no good
evidence that the Pythagoreans actually took this step. Heath
(*Arist.*, p. 115) cites Aëtius for the statement that the Pythagoreans
regarded the sun as spherical (οἱ Πυθαγόρειοι σφαιροειδῆ τὸν ἥλιον,
Aët. ii, 22, 5 = *Dox. Gr.*, p. 352); but this comes from Stobaeus'
Eclogae (compiled about AD 500), is unsupported by any other
source,[88] and cannot be regarded as reliable testimony for
Pythagorean beliefs. In fact, immediately above this passage,
Aëtius says that Alcmaeon (a Pythagorean of the early-fifth
century BC) believed that the sun was flat (πλάτυν εἶναι τὸν ἥλιον,
DK 24 A4). The same Alcmaeon, in agreement with 'certain of
the mathematicians' is also credited by Aëtius (*loc. cit.*) with
knowing that the planets have a movement from west to east.
This is the movement along the zodiac in the opposite direction
to the daily rotation of the heavens, and could be roughly accounted for on the Philolaic system by assuming different speeds of
rotation for each planet round the central fire (that is if the considerable deviations in latitude of the planets is ignored, as also
their retrograde movements and stationary points); but whether

knowledge of this double motion can be attributed to such an early figure as Alcmaeon is extremely doubtful. Aëtius also mentions (*DK* 58 B37c) three Pythagorean theories about the Milky Way: some thought that it was the track of a burnt-out star which had fallen from its proper place; others that it was the original course of the sun (presumably made visible by a sort of residual after-glow!); and others that it was the reflection of the rays of the sun in the heaven, just as the rainbow is its reflection in the clouds. Finally, we are told that they called the Pleiades the 'lyre of the Muses' and the planets the 'hounds of Persephone' (*DK* 58 c2).

Despite their insistence on the importance of number, we have practically no information about how the Pythagoreans applied numbers in their astronomical thinking. The sources tell us nothing, for example, about the periods allotted to the revolutions of the celestial bodies in the Philolaic scheme; it is true that later commentators indulge in much speculation about the different intervals and notes that they suppose to have comprised the harmony of the spheres, but they are not forthcoming about any actual parameters. Presumably the very fact that the scheme envisaged the revolutions of the different bodies as occurring in a particular order shows that some attention must have been paid to observed phenomena, and some quantitative information about solar, lunar, and planetary movements must have been utilized.

The only concrete evidence we have in this connection is a statement by Censorinus, a Roman grammarian of the third century AD who wrote a work, *De Die Natali*, containing a certain amount of calendaric information (not always accurate), that Philolaus the Pythagorean made a 'Great Year' consist of 59 years, including 21 intercalary months, and an ordinary year of $364\frac{1}{2}$ days (*op. cit.* 18, 8). A 'Great Year' was a period after which sun, moon, and planets were supposed to occupy exactly the same positions again as at the beginning;[89] but it came to be regarded as a common multiple of solar and lunar periods only for calendaric purposes (see below), and this is the sense in which Censorinus uses the expression – he gives several figures for this period as put forward by various authorities (*op. cit.* 18, 5ff.). On the basis of this attribution to Philolaus, the Italian scholar Schiaparelli[90] sets out a list of planetary periods (duly reproduced by Heath, *Arist.*, p. 102, note 2) which he compares with the modern figures, claiming that they show an extraordinarily close correspondence. It is important to realize that there is no evidence whatsoever that these figures were known to Philolaus; they are obtained merely by dividing the 21,505½ days of his Great Year by the nearest whole number of revolutions which each planet makes during that period according to modern knowledge, but counting Venus, Mercury, and the sun all together as completing one revolution in 364½ days. This foisting on an ancient scientist of knowledge based on modern data is a common feature of many present-day treatments of ancient astronomy, and one that it is continually necessary to guard against.

Another Pythagorean, Archytas, a contemporary and friend of Plato, and a mathematician of note who used a three-dimensional construction to solve the famous problem of doubling the cube,[91] provides some evidence of a general kind that mathematical techniques were to some extent applied to natural phenomena; he says (*DK* 47 B1), 'the mathematicians transmitted to us clear means of discerning (διάγνωσιν) about the speed of the stars and risings and settings, and about geometry and numbers and sphaeric [i.e. the geometry of the sphere with particular reference to astronomical problems] and not least about music.' This is the Archytas whom Horace apostrophizes as 'measurer of earth and sea and sand without number' and as one who has 'scaled the airy dwellings and traversed the round heavens with a mind that was after all to die'.[92] The concept of the spherical earth must soon have provoked attempts to estimate its size, and it is possible that a figure of 400,000 stades for the circumference, which Aristotle (*De Caelo* ii, 14, 298a15–17) attributes to 'some of the mathematicians', may have derived from Archytas (although it might equally well have come from Eudoxus – see below); and Horace's last two lines might refer to astronomical investigations by Archytas. He is also said (by Eudemus, *DK* 47 A24) to have asked the percipient question whether, if he stood on the outermost edge of the universe, i.e. the sphere of the fixed stars, he would be able to extend his arm and stick outwards; the natural affirmative answer would entail acceptance of a boundless universe.

Archelaus, a pupil of Anaxagoras, seems to have held much the same astronomical opinions as his master, but to have introduced a few variations. According to the tertiary sources, both regarded the stars as fiery masses, with the earth lying motionless at the centre of the cosmos, and the heavens tilted towards the south; but Archelaus apparently considered the earth not as a flat disc, but as a disc with a raised edge and a hollow middle part. This, he thought, explained why the sun does not rise and set at the same time in all regions, as it ought to do if the earth were level (ὁμαλή – DK 60 A4 = Hippol., *Ref.* i, 9, 4). Presumably the idea was that the sun would appear to rise and set at different times behind the raised outer rim according to the position of the inhabitants on the slopes of the hollow part; again this is a recognizable attempt to accommodate theory to the facts of observation (*cf.* p. 59). We are also told (*DK* 60 A1 and A4) that he regarded the sun as the largest of the celestial bodies, the moon as the next largest, while the others (not specified) were of various sizes.

Another philosopher who followed Anaxagorean notions was Diogenes of Apollonia (*flor. c.* 430 BC). For him, the ἀρχή was air, which he considered to be divine, eternal, immortal, and the guiding principle of the whole universe. From air, by the different combinations of its characteristics, i.e. temperature, moistness or dryness, rarefaction and condensation, and motion, were produced all the phenomena of the visible world. Warm air is the mark of intelligence, the amount of it contained in a particular organism determining its character. Pure divine air was fiery, and there were innumerable gradations of warmth down to inanimate objects which contain no warm air (fragments 5, 7, and 8 and the tertiary evidence *DK* 64 A5, 6 and 7). Much of this (amusingly satirized in Aristophanes' *Clouds* 225f.; 264; 828f.) recalls what the tertiary sources assert of Anaximenes. He also regarded air as the ἀρχή, and it is presumably recollection of this that caused part of the doxographical tradition to make the absurd assertion that Diogenes was actually a pupil of Anaximenes (*DK* 64 A1).

The scanty fragments of his work remaining give us no details about his astronomical ideas, for which we have to trust the tertiary sources. According to these he regarded the earth as 'round' (στρογγύλη), supported by air in the centre of the cosmos (*ibid.*). Unfortunately, στρογγύλος (like its English equivalent) is ambiguous in meaning;[93] it may mean 'round' like a ball (synonymous with σφαιροειδής, 'spherical'), or 'round' like a dish, or simply 'curved' (see *LSJ s.v.*). Here, since Simplicius tells us (*DK* 64 A5) that Diogenes followed Anaxagoras' views in many respects, the word can hardly mean 'spherical', but refers to a disc-shaped earth (*cf.* Guthrie, vol. ii, p. 372 note 1); he is also coupled with Anaxagoras in thinking that the cosmos was tilted towards the south (Aëtius, *DK* 59 A67). The stars, the sun, and the moon he regarded as consisting of red-hot pumice-stone (κισηροειδῆ), through the pores of which came rays from the aether. This sounds very odd, particularly for the moon, which Anaxagoras knew shone by reflected sunlight, and it is likely that the source, Aëtius (*DK* 64 A12–14), has his facts muddled – he even says that Diogenes considered that the sun was extinguished by cold acting in opposition to the heat! Another obvious borrowing from Anaxagoras is the notion that, as well as the visible stars, invisible stones are also carried round the earth like the one that fell at Aegospotami (A12). This refers to the fall of a large meteorite in 467 BC, which caused a considerable sensation and which was supposed to have been predicted by Anaxagoras (*cf. DK* 59 A11 and 12). To him we must also trace the idea that the drawing up of moisture by the sun's heat from the regions round the earth produces winds and the turnings of sun and moon; the remaining moisture forms the sea which will gradually become dried up (A17). The same process was used to explain the annual flooding of the Nile in summer, the sun drawing water from the sea and releasing it into the river (A18).

From what we know of the opinions of Empedocles, Anaxagoras, Archelaus, and Diogenes (and it must be remembered that we are largely dependent on the tertiary sources for them), it would seem that there was a common store of astronomical ideas in the second half of the fifth century BC which was drawn on by all the thinkers of the period.[94] Certainly, the atomists, Leucippus and Democritus, whom we come to next, drew on it. This is not the place to give a detailed account of the atomic theory developed by them, modified later by Epicurus, and expounded with missionary zeal by Lucretius in his *De Rerum Natura*. Suffice it to say that, according to this theory, matter consists of the conglomeration of many, small, invisible (and indivisible), homogeneous particles, which originally were scattered in infinite numbers throughout the void, differing from each other only in size and shape (some being round, some hooked, some triangular, and so on). By the action of the primeval vortex (δίνη), vaguely described as having been set in motion by necessity (δι᾿ἀνάγκην), the atoms came together, like shapes being attracted to like, to form first a sort of spherical membrane or caul (σύστημα σφαιροειδές . . . οἷον ὑμένα). As this whirled round, the finer atoms (e.g. of fire) went out towards the surrounding void and formed the celestial objects, while the coarser ones collected towards the centre and formed the earth and all its contents. This process goes on continuously throughout the limitless void, local conglomerations of atoms coming together to form countless worlds which grow, flourish, and then dissolve into their constituent atoms again (Diog. Laert. ix, 30ff. = DK 67 A1; Aëtius, A24).

Of the two proponents of atomism, Leucippus was the older, and his work, the 'Great World-System' (μέγας διάκοσμος), is generally dated to between 440 and 430 BC; Democritus was about ten years younger than Socrates (born in 469) and long outlived him. In the sources they are generally mentioned together when the basic principles of the system are being described, but their astronomical views are reported as showing considerable differences, and so here they are treated separately.

According to Diogenes Laertius (ix, 33), Leucippus said that the circle of the sun was the outermost, that of the moon the nearest to the earth, and those of the other celestial bodies (not specified) were in between. The earth is described as 'riding (or being carried) in a whirl round the centre' (τὴν γῆν ὀχεῖσθαι περὶ τὸ μέσον δινουμένην – ix, 30), a description which at first sight seems to show strong affinities with the Pythagorean moving earth.

However this is so foreign to what we know of the rest of atomist astronomy that it seems certain that Diogenes' account is confused here: he knew that the whole cosmos was supposed to have been formed by the action of the δίνη and carelessly uses the verb δινέω in connection with the earth as though the process was still going on. Other ideas attributed to Leucippus by Diogenes are that the earth is 'drum-shaped' (τυμπανώδης – cf. Aëtius, DK 67 A26), i.e. shaped like a tambourine, that the stars are made red-hot by the speed of their revolution, as is also the sun which is made more fiery by the stars, and that the moon receives only a small portion of fire. Then follows a passage where solar and lunar eclipses are apparently explained by the tilting of the earth towards the south, the northern part being always snowy, very cold, and frozen (ix, 33). The bizarreness, not to say incomprehensibility, of this 'explanation' has led most scholars to assume a lacuna in the text, because although the tilting might be adapted to explain differences in the seasons and the lengths of day and night (as had already been suggested by Anaxagoras, who makes the cosmos, not the earth, tilt – see above, p. 59), it is difficult to see how it can be made to account for eclipses. On the other hand, it is obvious that Leucippus' astronomical ideas were remarkably primitive for his time, and it may well be that he failed to understand the reasoning behind the theory of tilting. Diogenes ends his account by saying that, according to Leucippus, the sun is seldom eclipsed, while this is constantly (συνεχές) happening to the moon because their circles are unequal. Aëtius adds very little; he says that both Leucippus and Democritus considered the cosmos as spherical (DK 67 A22), and gives as a further reason for the tilting of the earth the porousness or rarity (ἀραιότης) of its southern regions (A27) – presumably the picture was of a top-heavy earth weighed down by snow and ice in the north and out of balance with the dry, warm, and light south.

Democritus' astronomical views are much more sophisticated than his predecessor's and show a greater willingness to pay attention to the facts of observation, but on the theoretical side he made no advances and his ideas are an amalgam of those of Anaxagoras and his pupils. In particular, he seems to have rejected or ignored the theoretical advances made by the Pythagoreans and taken up by Plato, e.g. the concept of the spherical earth. It was, perhaps, Plato's consciousness of the defects of Democritus' astronomical views that gave rise to the story, reported by Diogenes Laertius on the authority of Aristoxenus, that Plato wanted to burn all Democritus' books he could get hold of, but was dissuaded from this by two Pythagoreans (Diog. Laert. ix, 40).

Once again we have to rely on secondary and tertiary sources for the details of Democritus' astronomical beliefs, since of the nearly three hundred passages listed by DK as B fragments only half a dozen have any relevance at all to astronomy. Thus B5b, B11r, and B13, which can hardly count as actual fragments of Democritus' writings and might well have been included among the A references, are quotations of titles only, namely Περὶ τῶν πλανήτων (On the Planets), Μέγας ἐνιαυτὸς ἢ 'Αστρονομίη (the Great Year or the Astronomy) – also referred to as 'Αστρολογία and Περὶ ἀστρονομίας) – and Παράπηγμα (Calendar). Fragment B14 is a collection of data from the last-named, mostly taken from the calendar attached to Geminus' Isagoge and Ptolemy's Phaseis (on these see pp. 84f.). Fragment B15 is concerned with the shape of the inhabited world and is a very dubious fragment anyway (see below). Fragment B25 informs us that Democritus regarded ambrosia as the vapours by which the sun is nourished.[95] To judge by the list of over sixty titles attributed to him by Diogenes Laertius on the authority of Thrasyllus (an Alexandrian astrologer of the late-first century BC), Democritus' intellectual interests extended over a wide range, and it is a great pity that not a single one of his writings has come down to us complete.

In astronomy, he corrected Leucippus' erroneous view of the order of the celestial bodies and stated that the moon was closest to the earth (which he regarded as having been formed before the stars). Then came the sun, and then the planets (unspecified) which have not all the same height (i.e. distance from the earth) – this is according to Hippolytus (DK 68 A40). Aëtius says that he put the fixed stars first (reckoning inwards from the surrounding void), then the planets (again unspecified), then the sun, then

Φωσφόρος (i.e. Venus), and then the moon. The fixed stars he regarded as stones (presumably fiery) and the sun as a red-hot stone (A85–7). On the other hand, Seneca (*Nat. Quaest.* vii, 3, 2) asserts that Democritus suspected that there were several stars which were not fixed, but he did not enumerate them by name because the courses of the five planets were not yet understood.[96] If this were true, and it may very well be, granted Democritus' apparent rejection or ignorance of Pythagorean astronomical ideas (despite a misguided attempt by part of the doxographical tradition to connect him with both Pythagoras himself and Philolaus – *cf.* Diog. Laert. ix, 38), then it is difficult to see what he could have put into a work entitled *On the Planets.*

Sun and moon he regarded as composed of smooth and round atoms, λείων καὶ περιφερῶν ὄγκων (ix, 44); περιφερής describes spherical atoms which are the constituents alike of fire and the soul substance and νοῦς 'mind', which he equated with the divine (Aëtius, A74, νοῦν τὸν θεὸν ἐν πυρὶ σφαιροειδεῖ: A135 = Theophrastus, *De Sensu* 68, τοῦ θερμοῦ τὸ σχῆμα σφαιροειδές). Aristotle, who wrote a book on Democritus (*cf.* Simplicius, A37), tells us that he regarded the sphere as the most mobile shape (A101 = *De Anima* i, 2, 405a11). Originally, sun and moon were not fiery, being made of the same type of atoms as constituted the earth, but later, in the process that led to the enlarging of the sun's circle, fire was cut off in it (A39). This view did not prevent Democritus from realizing that the moon shines by light from the sun (Plutarch, A89a).

Despite his avowed dislike of Anaxagoras, whom he accused of plagiarizing old ideas about the sun and moon (B5), many of Democritus' own astronomical ideas, as reported in our sources, are identical with those of Anaxagoras. Thus both believed that the moon was like the earth in having mountains and glens and plains (DK 59 A77 and 68 A90), and both gave the same explanation of the Milky Way (68 A91) and of comets (68 A92); both regarded the earth as flat, but Democritus followed Archelaus in supposing that it was disc-shaped and hollow in the middle (A94). Water collected in the hollow parts, and local excess accumulations of water caused the land mass to shift and so produced earthquakes (A97). This passage comes from Aristotle, who, as we have seen (p. 46), connects Democritus with Anaximenes and Anaxagoras in believing the earth to cover the air beneath it like a lid. The earth was tilted towards the south not, as Leucippus had thought, because its northern parts were heavier, but because the southern part through an over-abundance of natural produce and growth outweighs the virgin north (τὰ βόρεια ἄκρατα); either Aëtius' account is faulty (A96) or Democritus had very confused ideas of the Anaxagorean notion of tilting, since the passage actually talks about the 'greater weakness of the southern part of the surrounding (air)', διὰ τὸ ἀσθενέστερον εἶναι τὸ μεσημβρινὸν τοῦ περιέχοντος, which makes very little sense here.

In the beginning the earth moved about (πλάζεσθαι) because of its smallness and lightness, but as it grew denser and heavier it remained stationary (A95). According to Agathemerus (B15), compiler of a small and very bad geographical treatise of uncertain date but undoubtedly post-second century AD, Democritus thought that the inhabited part of the earth (ἡ οἰκουμένη) was not round (στρογγύλος again), as the ancients believed, with Greece in the centre and Delphi in the middle of Greece. This was the traditional picture, descended from the Homeric concept of the all-encircling Ocean stream, and already criticized by Herodotus (iv, 36). Democritus, on the other hand, maintained that it was oblong (προμήκης), its length (i.e. west to east) being half as long again as its breadth (north to south). The same source informs us that Democritus wrote a Γῆς περίοδος, 'Circuit of the Earth', and a Περίπλους, literally a 'sailing round', that is a navigational account; but neither of these titles occurs in the list given by Diogenes Laertius (A33). Agathemerus is a poor authority, and it is surprising that Diels should classify the passage as a B fragment.

Finally, Lucretius tells us (De Rerum Natura v, 621f. = A88) that Democritus believed that the sphere of the fixed stars revolved with the greatest velocity, while the sun and moon, being nearer the earth, were less affected by the 'revolution of the heavens' (caeli turbo) and therefore moved more slowly, the moon being the slowest of all; hence the sun was overtaken by the zodiacal signs in a year and the moon in a month, but to our eyes it seemed as though the latter moved faster. This is an idea that is implicitly criticized by Plato in the *Laws* (see below, p. 139; *cf.* Aristotle, *De Caelo* ii, 10). It is evidently an attempt to account for the different speeds of revolution of sun and moon relative to the fixed stars from west to east, but it manifestly confuses the diurnal revolution of the whole heaven in the plane of the equator with the movements of the sun and moon in the plane of the ecliptic. If it were merely a question of relative speeds, all the revolutions would be in the same plane. Alexander of Aphrodisias, a third-century AD commentator on Aristotle, attributes to the Pythagoreans the idea that the celestial bodies which are furthest away move with greater speed than those closer to the earth (*in Met.* A 5); but this cannot be true of the Philolaic system, in which (as we have seen) the earth itself was a moving body and the sphere of the fixed stars was either motionless or was endowed with a very slow movement.

In the last decades of the fifth century BC, the Hesiodic type of observational material mentioned above (p. 60) began to be correlated with current astronomical ideas, and one of the first results was the emergence of the 'parapegmata' or astronomical calendars. These were a kind of almanac engraved originally on stone or wooden tablets, giving astronomical and meteorological phenomena for all the days of the month. Each day was originally represented by a hole at the side of that part of the engraved text which gave the prognostication for that particular day, and into the hole a movable peg was inserted; next day the peg was inserted into the next hole and so on (hence the name, from παραπήγνυμι, 'fix beside'). Fragments of four stone parapegmata from Miletus and elsewhere, the earliest of which dates from the second century BC (thus comparatively *late* in the development of Greek astronomy), have actually been found.[97] The information given was of the type: 'Day 6: the Pleiades set in the morning; it is winter and rainy' or 'Day 26: summer solstice; Orion rises in the morning; a south wind blows'; and apparently from the first such data were also written up as separate texts.[98]

It is from two extant examples of these, the calendar attached to Geminus' *Isagoge*[99] and Ptolemy's *Phaseis*,[100] that we derive most of our information about them.[101] Both these calendars were compiled at a time when mathematical astronomy had already reached a high level of proficiency, and so they cannot be regarded as typical of the earliest examples; in fact, they contain meteorological and astronomical data[102] deriving from many different observers in different localities. Thus the calendar attached to Geminus' work uses as sources Democritus (the earliest authority cited, which confirms the title, Παράπηγμα, attributed to him – see p. 81), Euctemon, Meton, Eudoxus, and Callippus. Ptolemy (about three hundred years later) adds ἐπισημασίαι from the Egyptians (who observed παρ'ἡμῖν, i.e. in Alexandria – exactly who these were is not clear), Hipparchus, Julius Caesar, Dositheus, Metrodorus, Philippus, and Conon. Ptolemy, of course, was aware that the data connected with the various days of the year depended on the latitude of the observer (cf. pp. 14f.), and he gives (Phas., p. 67 ed. Heib.) the places where the observers made their observations. This fact was apparently not appreciated by the general public who presumably used the parapegmata as calendaric guides for the business of everyday life.

Thus we find a body of material that soon became traditional in character and was copied and recopied indiscriminately by compilers and popularizers who never made any observations themselves and ignored the discrepancies arising from the different regions to which the data were applicable. One result of this can be seen in the rustic calendar set out in ch. 25–31 of Book xviii of Pliny's Naturalis Historia (first century AD), where he complains of the different dates he finds given for the same phenomena in various authorities, and demonstrates his own scientific incompetence in dealing with the material.

Obviously, to be any use at all such data must be attached to a fixed calendaric scheme based on the solar year, and here a well-known difficulty obtrudes itself. The most natural means of dividing the year into convenient periods longer than single days and nights is by the lunar (synodic) month – the time that elapses between one phase of the moon and the next recurrence of that phase. The Babylonian calendar, for example, like the early

Jewish calendar and the official Mahommedan calendar to this day, was at all times a lunar calendar; a new month began with the first visible appearance of the crescent of the new moon (hence the importance attached to being able to predict this and the elaborate computational schemes for this purpose).[103] Moreover, the moon plays a very important role in the religious practices of most peoples, in that the occurrence of religious festivals is largely regulated by its phases (the date of Easter, for example). Nilsson has shown that most Greek festivals took place at or near full moon (*Primitive Time-Reckoning*, p. 343), and that all the Greek names for months (which vary from city to city, each having its own particular calendar) are connected with religious festivals.[104]

Unfortunately, the lunar month is incommensurable with the solar year: no convenient whole number of months makes up exactly one year. The lunar month is a little more than $29\frac{1}{2}$ days; 12 of these amount to 354 days and 13 to 384 days, whereas the sun takes very nearly $365\frac{1}{4}$ days to complete one full cycle. Thus a purely lunar-based calendar is very soon going to become greatly out of step with the sun. A festival supposed to be held at full moon in mid-summer would in the course of years be taking place in the autumn, winter, or spring. In fact, the Mahommedan year of 12 lunar months is about 11 days out by the sun each year, and every date in this calendar goes the complete round of the seasons every 33 years.

Now this is all very well for the followers of Mahommet, but it would not do for the Greeks. One of the characteristics of both Greek and Roman religion is the insistence on exact ritual; the gods were displeased if the rites were not carried out in exactly the fashion laid down, and this, of course, included having them on the same day or days each year. Geminus, *Isagoge* ch. 8, the *locus classicus* for Greek calendaric cycles, is very clear on this point. There is also the obvious absurdity in holding, say, a harvest thanksgiving festival at a time when the corn had not even been planted. So the Greeks expended considerable thought in establishing a luni-solar year, in which the months and days are measured by the phases of the moon but which still keeps in step with the sun. The problem then is to find an extended period which is a common multiple of the lunar and solar cycles, and to intercalate months as necessary during any one period.

According to Geminus (*loc. cit.*), the earliest such intercalation cycle was one of 8 years, the 'octaëteris', since 8 solar years of 365 days are roughly equal to 99 lunar months of $29\frac{1}{2}$ days. However his description of how this evolved is unsatisfactory (see below, pp. 188–89) and there is considerable doubt concerning the origin of this cycle. Censorinus (*De Die Nat.* 18, 5) tells us that it was usually ascribed to Eudoxus, but that other names were also connected with it, including that of Cleostratus of Tenedos. This latter is a shadowy figure of the second half of the sixth century BC, mentioned less than a dozen times in the tertiary sources (see *DK* 6) and supposed to have written on astronomy. The scholiast on Euripides, *Rhesus* 528 quotes two of his hexameters (Diels supplies a third, *DK* 6 B1), and it is possible that he wrote a poem in the Hesiodic manner giving some information about the constellations then known; but it is wholly impossible that he introduced the zodiacal signs and understood the concept of the ecliptic, as Pliny tries to make out, since this does not appear until the end of the fifth century BC;[105] and attempts to build up Cleostratus as a key figure in early Greek astronomy[106] are certainly misconceived.

The first well-attested intercalation cycle is connected with the names of two Athenian astronomers, Meton and Euctemon, who flourished about 430 BC and, according to Ptolemy (*Phas.*, p. 67, 2), made observations at Athens, in the Cyclades, in Macedonia, and in Thrace. In the *Almagest* (iii, 1) they are cited for observations of the summer solstice, including one on 27 June 432 BC, which, despite their inaccuracy by the standards of later astronomy (emphasized by both Hipparchus and Ptolemy – modern calculations show that the solstice actually occurred about $1\frac{1}{2}$ days later), were nevertheless used as confirmation of the figure that Hipparchus decided on for the length of the year and that Ptolemy also accepts (viz. $365\frac{1}{4}$ days less $\frac{1}{300}$ of a day). Meton suggested a period of 19 years, called after him the Metonic cycle, to bring the lunar month into correlation with the solar year. This cycle contained 235 lunar months (7 of which were intercalary) and 6,940 days, and, according to Geminus,[107] 110 of the months were 'hollow', i.e. of 29 days each, and 125 'full', i.e. of 30 days each. This would give a mean lunar month less than two minutes too long and a solar year of $365\frac{5}{19}$ days, about 30 minutes too long.[108]

Meton and Euctemon are frequently cited in the parapegmata, and are the earliest names connected with the observations of equinoxes as well as solstices. Moreover, according to a second-century BC papyrus fragment known as the *Ars Eudoxi*,[109] Euctemon gave the lengths of the astronomical seasons starting from the summer solstice as 90, 90, 92, and 93 days respectively (the modern figures to the nearest whole day are 92, 89, 90, and 94), which shows that he was aware of the non-uniformity of the sun's course round the earth. Now, recognition of the equinoxes and of the inequality of the seasons implies a comparatively sophisticated stage in astronomical thought, and presupposes at least some knowledge of the concept of the celestial sphere and a spherical earth – see my article in *JHS* 86, 1966. Thus the Pythagorean ideas were now beginning to bear fruit when applied to the observational material that was available, and it would seem that a much clearer picture was being obtained of at least the sun's course, with the solstices and equinoxes marking the four seasons of the solar year.

An essential part of this picture is the concept of the sun's circuit of the heavens marked by its passage through the zodiacal constellations in a plane *inclined* to that of the equator. As we have seen, this was not part of the Philolaic scheme. The invention of the ecliptic as the sun's oblique path is attributed by Eudemus to Oinopides of Chios (*DK* 41, 7), who we are told was a little younger than Anaxagoras (*ibid.* 1), and who is mentioned in the pseudo-Platonic dialogue *Amatores* (132a–b) in connection with drawings of inclined circles; he, too, seems to have been a Pythagorean. No actual figure for the obliquity of the ecliptic is ascribed to Oinopides, but he may have known the rough estimate of 24° (see below, pp. 157–58). According to Censorinus (*De Die Nat.* 19, 2) Oinopides made the length of the year $365\frac{22}{59}$ days, while Aelian and Aëtius ascribe to him a Great Year of

59 years (*DK* 41, 9). It is possible, as Tannery suggests,[110] that, starting with the figures of $29\frac{1}{2}$ days for the lunar month and 365 days for the solar year, Oinopides realized that the smallest whole number of years to contain a whole number of lunar months would be 59 ($2 \times 29\frac{1}{2}$), containing 730 (2×365) months which would be equivalent to 21,557 days, and this divided by 59 would give $365\frac{22}{59}$ days.

Equinoxes and solstices and other parapegma data are also mentioned in the medical treatises of the Hippocratic corpus, especially the works entitled *On Airs, Waters, Places* and *On Diet*.[111] These treatises are notoriously difficult to date,[112] but certainly none of them can be earlier than 425 BC.

Hence all the evidence points to the conclusion that the last decades of the fifth century BC saw the real beginning of mathematical astronomy in Greece, although the celestial sphere as a fully developed concept does not appear until the fourth century. It seems certain that calendaric problems provided the initial impetus for this development. There is ample evidence that in the fifth century the Athenian civil calendar was in a state of confusion. Aristophanes in the *Clouds* (610*ff.*) makes the moon complain that the Athenians were not arranging the days properly in accordance with its phases (*cf. Peace* 406). Thucydides (v, 20) explains that time-reckoning by the tenure of office of state magistrates (this was the normal official method, e.g. an event would be dated in such-and-such a year of so-and-so's archonship) was bound to be inaccurate (it would anyway only be readily intelligible to a reader in one city, since each city had its own magistrate list and calendar), and the only safe method was by counting summers and winters. This would be with the help of a parapegma. He also uses the astronomical phenomena listed in the parapegmata, such as the solstices (vii, 16; viii, 39) and the rising of Arcturus (ii, 78) for dating specific events. It must be realized that astronomically based cycles, such as the Metonic cycle, were only used in scientific texts, while in the ordinary civil calendar of each state no systematic scheme of intercalation was apparently in use, but intercalation depended on the vagaries of officialdom.[113]

The new astronomical ideas naturally did not immediately

win full acceptance, and many of the old, crude notions of the early Pre-Socratics lingered on into the fourth century and even later, to judge from the sources. Antiphon (the sophist of the second half of the fifth century, not the orator), who was no mean mathematician,[114] is said to have believed that the moon shone by its own light, but that the concealed part round it was dimmed by the proximity of the sun, since the stronger fire dims the lesser one, which happens also with the other stars (*DK* 87 B27). He also thought (B26) that the sun's fire fed on the damp air round the earth and that its risings and settings were caused by the recurrent failure of its burning owing to the opposing effect of moisture; and he is coupled with Alcmaeon and Heraclitus in supposing that the moon was bowl-shaped (σκαφοειδής), and that eclipses were caused by the turnings and inclinations of its bowl (B28) – all three of these 'B' fragments (in Diels' classification) come from Aëtius. Metrodorus of Chios, one of Democritus' disciples, apparently still believed that night and day were caused by the alternate extinguishing and rekindling of fiery vapour in the sun, that this also produced eclipses, and that the sun created the stars out of 'radiant water' (λαμπροῦ ὕδατος, *DK* 70 A4). Aëtius connects him with Anaximander and Crates (*flor.* second century BC) in putting the sun as the highest of the celestial bodies, then the moon, and below these the fixed stars (which were illuminated by the sun) and the planets (*DK* 12 A18; *cf.* 70 A9). The same source tells us that according to both Thales and Metrodorus the moon was illuminated by the sun (70 A12 – the coupling together of these names is patently absurd), and that Metrodorus explained the Milky Way as the sun's circle. Hecataeus of Teos or its colony Abdera is joined with Heraclitus by Aëtius in allegedly stating that the sun is an 'intelligent, ignited mass from the sea' (ἄναμμα νοερὸν τὸ ἐκ θαλάττης, *DK* 73 B9); Hecataeus is dated to the end of the fourth century BC.

The sophists in general did not profess a specialist knowledge of astronomy – their forte was to teach people how to 'get on in life' by inculcating the arts of rhetoric and argumentation and the handling of affairs. However Hippias, a sophist who was specially disliked by Plato, is portrayed as claiming a profound knowledge of the stars and celestial matters (*Hipp. Maj.* 285*b*), and certainly seems to have been an accomplished mathematician.[115] Unfortunately, there is no evidence as to what his astronomical ideas were. According to the pseudo-Plutarchian *Lives of Ten Orators*, the funeral monument of Isocrates, who died in 338 BC, depicted Gorgias, one of the older sophists (*c.* 485–375 BC), as looking at an astronomical sphere (εἰς σφαῖραν ἀστρολογικὴν βλέποντα – *DK* 82 A17); but this is probably no more than artistic convention, since a globe was a recognized decorative feature.[116] There is no evidence in the extant fragments of Gorgias or in the tertiary sources[117] that he paid any particular attention to astronomy.

CHAPTER V

PLATO

WITH THE TURN of the fifth century BC, the next great name we meet is that of Plato (428–347). Plato's importance in the development of Greek astronomy is easy to underestimate; he was not a practising astronomer (no observations of his are recorded), he made no original contributions to astronomical theory, and he advocated the study of astronomy in a way that has lent itself readily to misinterpretation (see below on *Rep.* vii, 530*b–c*). Nevertheless, it is hardly too much to say that, had it not been for Plato, Greek astronomy might never have become the influential force it did – with incalculable consequences for medieval and later science.

It was his particular concern with the place of mathematics and astronomy in education – not only the education of the guardians, the highest ruling class in his ideal state, but the general education of the ordinary citizens as well – that makes him such an important figure in the history of science. For although the sciences are in some measure defective in his eyes because they begin from unexplained hypotheses (*Rep.* 533*b–d*), nonetheless they are of very great value in drawing the soul towards 'the Idea of the Good' (πρὸς τὴν τοῦ ἀγαθοῦ ἰδέαν, *id.* 526*e*) and in preparing it for the understanding of the only means by which such knowledge can be attained, namely, the method of dialectic, which is the real objective towards which the ideal education should be directed (*id.* 531*c*; 532*a–c*).

If we read in chronological order those portions of the dialogues that discuss the role and place of astronomy in education, we can hardly fail to be struck by the gradually increasing emphasis that Plato laid on astronomical studies as he grew older. In the *Republic* (vii) it is one of the four branches of mathematics, namely, arithmetic, geometry, solid geometry, and astronomy, listed in their logical order of development, and not necessarily of more intrinsic importance than the others. In the *Laws* (which belongs to the last thirteen years of Plato's life) astronomy is classified as one of the three subjects (together with arithmetic and mensuration) about which *everyone* ought to know a little, not just the guardians and the special classes who might be expected to derive professional benefit from it. The secondary education of the citizens (i.e. up to sixteen years old) must include enough astronomy to enable them to understand calendaric problems for the proper regulation of the city's life (809c–d; 817e–818a). Moreover, astronomy is now regarded as one of the two main ingredients in the education of the most important body in the state, the 'nocturnal council' (νυκτερινὸς σύλλογος, the standing committee of the guardians, 951d–e; 961a–b), for inculcating a proper belief in and reverence for the gods, and Plato goes to some lengths to absolve those studying it from the charge of atheism (966d–967d). Finally, in the *Epinomis* (which there is no compelling reason to think is not Platonic), οὐρανός is regarded as the greatest god and responsible for all benefits to mankind including ἀριθμός, 'number', whereby it teaches man to count by providing sun and moon for the calculation of days, months, and years (977a–979a). Furthermore, it is now stated explicitly that the one subject of prime importance for giving the best education is ἀστρονομία, since your true astronomer must needs be σοφώτατος (990a–c).

Any attempt to elucidate the details of Plato's astronomical beliefs faces at the outset two well-known difficulties: (*a*) the difficulty of deciding whether the opinions put into the mouth of a particular speaker in a dialogue can legitimately be regarded as those of Plato himself;[118] and (*b*) the fact that the context of both the passages where any astronomical details are given makes it clear that the account is at best provisional and certainly not intended to be definitive.[119] In what follows I make the assumption that, in his cosmological descriptions, even when he is telling a story or describing a system which does not pretend to be more than a probability, Plato is unlikely to have introduced deliberate falsifications, or set down opinions which, with due allowance being made for poetic imagery and dramatic effect, he would not himself have supported – at least at the time of writing. I also assume that, in the absence of any clear indications to the contrary, the views that he makes his chief characters express, be they Socrates, Timaeus, or the Athenian Stranger, represent Plato's own views on the subject in question – which, of course, did not necessarily remain the same throughout his long life.

In the *Phaedo* (perhaps written when Plato was in his thirties), as a prelude to the final proof of the immortality and indestructibility of the soul, Socrates is made to give a brief rèsumè of his intellectual quest for knowledge about matters pertaining to 'physics' (περὶ φύσεως ἱστορία, 96a). He describes how after much fruitless search which merely left him more confused than ever,[120] he heard someone reading from a book of Anaxagoras in which Mind (νοῦς) was posited as the governing principle of the universe. He was encouraged by this, since obviously Mind would have seen to it that the universe was arranged in the best possible manner and for the best possible reasons. This is that teleological strain in Plato's thought which comes out clearly in the *Timaeus*, where time and again the desire of the Demiurge to make his creation as perfect as possible is emphasized – e.g. 30a–b; 30d; 33b; 34b; 37 *et al.* Socrates eagerly read the rest of Anaxagoras' books, hoping to find the answers to such questions as whether the earth is flat (πλατεῖα) or round (στρογγύλη), whether it is at the centre of the universe and why, and to find information about the sun 'and the moon and the other stars and their speeds relative to each other and their turnings (τροπῶν) and other characteristics' (97d–98a). This may be compared with *Gorgias* 451c, where Socrates states that the λόγος of astronomy concerns 'the revolution of the stars and of the sun and of the moon, and their speeds relative to each other', τὴν τῶν ἄστρων φορὰν καὶ ἡλίου καὶ σελήνης, πῶς πρὸς ἄλληλα τάχους ἔχει.

His hopes, however, were disappointed because Anaxagoras did not in fact use his concept of Mind in the manner in which Plato himself was later to use the Demiurge, but postulated mechanistic causes such as 'airs and aethers and waters and many other absurdities' (ἄλλα πολλὰ καὶ ἄτοπα, 98c) to account for astronomical phenomena (see above, p. 57); but things like this, Socrates complains, cannot be regarded as the real cause (τὸ αἴτιον τῷ ὄντι, 99b), and people who think in this way are misusing the term 'cause'. The same applies to those who suppose that a vortex (δίνη – the reference is to Empedocles) surrounds the earth, which remains in position by virtue of the heavens, and those who envisage a flat earth like a kneading trough supported on a basis of air (according to Aristotle, De Caelo 294b14, this was a notion common to Anaximenes, Anaxagoras, and Democritus).

The topics for which Socrates is represented as consulting Anaxagoras' books are just those which might be expected to be matters of debate in the middle of the fifth century when he was a young man, and Pythagorean ideas about the sphericity of the earth and its position relative to the other celestial objects were being promulgated; the reference to relative speeds (τάχους τε πέρι πρὸς ἄλληλα) would be particularly apposite in connection with the Philolaic system (see above). Hence it is likely that the picture presented of a vain search for physical knowledge is intended to apply to the historical Socrates himself. On the other hand, we have seen (p. 90) that several of the crude astronomical notions of the Pre-Socratics lingered on well into the fourth century, when Plato was actually writing, so that to some extent such an intellectual quest might well have been Plato's own experience.

Later in the Phaedo, in response to a question by Simmias, Socrates outlines his views on the configuration of the earth, which, he has been 'convinced by someone' (ὡς ἐγὼ ὑπό τινος πέπεισμαι, 108c7), is very different from what is commonly supposed. He begins by asserting that if the earth is a circular body in the middle of the heavens (εἰ ἔστιν ἐν μέσῳ τῷ οὐρανῷ περιφερὴς οὖσα, 108e) there is no need to postulate anything to support it, because it will keep its position by reason of its 'equiformity' (ὁμοιότης) and 'equilibrium' (ἰσορροπία); this was an Ionian notion that, according to Aristotle (De Caelo 295b11–12), went back to Anaximander (see above). Judging from the thrice repeated and emphatic πέπεισμαι ('I have been convinced' – 108e1; e4; 109a7; as well as the earlier instance cited above), we may suppose that this part of Socrates' account at least is meant to be taken as a statement of accepted fact. To this, Simmias gives a ready assent (καὶ ὀρθῶς γε, 'quite right!' 109a8).[121]

Socrates then goes on to give a description of the earth, the gist of which is as follows. The earth is very large, and we inhabit only a small portion of it stretching from the Pillars of Heracles (Straits of Gibraltar) to Phasis (at the eastern end of the Mediterranean), dwelling like frogs or ants round a lake. There are many other inhabited hollows round the earth into which water, mist, and air have gathered, but the real surface of the earth is situated under the pure heaven (where the stars are) which is commonly called the aether. We think that we are living on the upper surface of the earth and call heaven what is actually only the air that has flowed into our hollow, in which we see the stars as fish would see them in and through water; our human imperfections preclude us from ever viewing the real heavens (109a9–110b).

Here Socrates emphasizes that he is telling a story, though that is worth hearing (εἰ γὰρ δὴ καὶ μῦθον λέγειν καλόν, ἄξιον ἀκοῦσαι, 110b1), and continues by saying that the earth as seen from above looks like a twelve-panelled sphere[122] of different colours, all much brighter and purer than those we know. Similarly, all the productions of nature generated in the many hollows on the outer surface of the earth are much finer and more perfect than ours. The inhabitants, some dwelling on the mainland, others round air just as we dwell round the sea, and others in islands surrounded by air, are as much superior to us in every sense as air is to water and as aether is to air; and they have the felicity of seeing the sun, moon, and stars as they really are (110b5–111c3). There are many hollows of different sizes and depths in the earth, itself honeycombed with subterranean passages which connect the hollows and through which flow many rivers of hot and cold water, of fire, and of mud. The whole complex is kept in constant motion by a sort of pulsation or 'see-saw' (αἰώρα, 111e4–5) at the centre of the earth, which is the region of Tartarus (Hades), conceived of as a great tunnel bored diametrically through the earth (111c4–112e3). Socrates then describes the chief (mythical) rivers, starting with Oceanus, which is the largest and outermost and flows round in a circle (ἐξωτάτω ῥέον περὶ κύκλῳ, 112e6); the others are Acheron, Pyriphlegethon, Styx, and Cocytus. Finally, we have the moving description of the fate meted out to the dead according to the quality of the lives they have led on earth, and hints of the joys in store for 'those who have sufficiently purified themselves through philosophy' (114c).

Now all this is avowedly a mythical story, and Plato makes Socrates specifically state: 'No man of sense should affirm decisively that all this is exactly as I have described it' (114d1-2, Bluck's translation); but this is not to say that a man of sense should automatically reject every idea in it as unscientific merely because it occurs in the context of a myth. As Stewart remarks,[123] 'The true object of the *Phaedo* myth is, indeed, moral and religious, not in any way scientific. . . . This moral and religious object, however, is served best, if the history or spectacle, though carefully presented as a creation of fancy, is not made too fantastical, but is kept at least consistent with "modern science" [i.e. the science of Plato's time]'. He was a consummate master of the use of myth in his dialogues, and, as such, well knew the necessity for keeping his creations within the bounds of probability without flagrantly violating the scientific preconceptions of his contemporaries.[124] The *details* of his descriptions are not necessarily to be pressed,[125] but it is legitimate to expect the underlying principles to conform to his own views on scientific matters; the alternative, of supposing that he would deliberately choose to give a false picture which in no way represented his own beliefs, is impossible for any 'man of sense' trying to understand Plato.

Thus it is easy to show that the islands surrounded by air, where superior human beings live disease-free lives and enjoy natural bounties more perfect than ours, are a variation on Pindar's Isles of the Blest (*Ol.* ii, 71 Bowra), a recurring theme in Greek literature, going back to Hesiod (*WD* 167ff.). The underground water system derives from the Sicilian Empedocles (Sicily is actually mentioned at 111e1), the pulsation of the waters is connected with theories of respiration and the circulation of the blood (see Burnet *ad loc.*), and the Homeric borrowings in the description of Tartarus and Oceanus are obvious. All this may be dismissed as

poetic colouring, not intended to give Plato's own scientific
opinions, although it might equally well be used as evidence for
his familiarity with certain currents of thought in the late-fifth
century. What cannot be dismissed is the basic concept underlying
the whole picture, namely, that of a spherical earth at the centre
of the universe. It seems to me that no unbiased reader of these
sections of the *Phaedo* can properly doubt that Plato had firmly in
mind the earth as a sphere, even though this is not actually stated
in so many words, and the idea that he envisaged its being any
other shape is completely untenable.[126]

Apart from Socrates' conditional statement of its roundness and
central position (see above), and the assertion that, from above,
it looks like a twelve-panelled sphere (ὥσπερ αἱ δωδεκάσκυτοι
σφαῖραι, 110*b*6 – this by itself should be enough to discount the
theory that Plato adhered to the Ionian view of the earth as a disc),
there are numerous passages where the spherical shape is implied.
In 111*c*, the other hollows on the earth's surface are described as
lying 'in a circle round the whole' (κύκλῳ περὶ ὅλην). In 112*d*, the
rivers of the earth's interior are pictured as 'going round in a
circle either once or even coiled many times round the earth like
snakes' (κύκλῳ περιελθόντα ἢ ἅπαξ ἢ καὶ πλεονάκις περιελιχθέντα
κερὶ τὴν γῆν ὥσπερ οἱ ὄφεις). In 113*b*, the River Pyriphlegethon,
after proceeding 'in a circle' (κύκλῳ), 'coils round within the
earth' (περιελιττόμενος τῇ γῇ – Burnet), and 'having coiled many
times below ground' (περιελιχθεὶς δὲ πολλάκις ὑπὸ γῆς) flows into
Tartarus. Similarly, in 113*c*, the Styx 'coils round' (περιελιττόμενος)
and 'goes round in a circle' (κύκλῳ περιελθών). Judged in isolation,
the last two examples might perhaps be taken as referring only to
circular motion in an earth of unspecified shape, but the cumu-
lative effect of the repeated use of κύκλος and περί (particularly
with γῆ) leaves no room for doubt that Plato was thinking of a
spherical earth; possibly the reason why Socrates does not affirm
it directly is that he knew (as did everyone else) that it was
accepted Pythagorean doctrine which there was no reason to
emphasize for Pythagorean hearers. Similarly, in the passage in the
Timaeus where Plato discusses the propriety of using such terms
as 'above' and 'below' in relation to the spherical universe

(62d–63d), the argument is hardly intelligible without the implicit assumption of a spherical earth at the centre (cf. Cornford ad loc.).

It is in the seventh book of the *Republic* (written perhaps c. 387, but set in the Athens of 421) that we find the fullest exposition of Plato's attitude to astronomical studies and their place in the best type of education. The actual details of the physical system he envisaged at this time are, however, heavily disguised in mythical language (forming part of the myth of Er in Book x) which, to a greater extent than in the *Phaedo* myth, is governed by considerations of poetic imagery and imaginative fancy. Hence the *caveat* entered above, about the danger of pressing the details, needs even more emphasis in this case.

One feature that receives more recognition in the *Republic* is the practical utility of astronomical knowledge in the business of life. Socrates in the *Phaedo* is represented as being so disgusted at his failure to learn anything of value about natural science from Anaxagoras' books that he turned to other matters (99e–100a). However in the *Republic* he expresses a qualified approval of the practical advantages of a knowledge of the mathematical sciences, although insisting strongly that their theoretical value is much more important (see below). In *Rep.* vi, 488d, the true pilot is depicted as necessarily having to make a study 'of the year and the seasons and the sky and the stars' as well as the winds and all the technical details of his craft. Similarly, the sun's importance is stressed as providing the seasons and the years and 'in a way' being the cause of everything in the phenomenal world (τρόπον τινὰ πάντων αἴτιος, 516c), since it is the origin of 'birth and growth and nourishment' (509b). In the simile of the Cave that opens Book vii, the prisoner who has been liberated from the realm of shadows, after first having to accustom his eyes to the unwonted brightness by observing initially only the shadows and reflections of things, and the heavens only by the light of the stars and the moon, is depicted as at last being able to look on the sun itself, 'not images of it in water or in some other material' (οὐκ ἐν ὕδασιν οὐδ' ἐν ἀλλοτρίᾳ ἕδρᾳ φαντάσματα αὐτοῦ, 516b). This, of course, is not to be taken literally; Plato was well aware of the dangers of gazing at the sun with the naked eye, and in *Phaedo* 99d expressly draws attention to the occupational risks run by those who observe eclipses of the sun, some of whom have their eyes ruined 'if they do not observe its reflection in water or something similar' (ἐὰν μὴ ἐν ὕδατι ἤ τινι τοιούτῳ σκοπῶνται τὴν εἰκόνα αὐτοῦ – cf. Laws 897d).[127] The description of the liberated man as being now able to 'look on the actual sun itself in its proper place and observe its nature' (αὐτὸν καθ'αὐτὸν ἐν τῇ αὐτοῦ χώρᾳ δύναιτ'ἂν κατιδεῖν καὶ θεάσασθαι οἷός ἐστιν, 516b) is a picturesque exaggeration, designed in a typically Platonic manner to emphasize the dramatic contrast between the former shadow life and the new life in the bright clarity of the upper world. We shall see later that a similar dramatic exaggeration has been the cause of a complete misunderstanding of Plato's attitude to astronomy.

In the well-known illustration of the Divided Line at the end of Book vi, Plato classifies the objects of knowledge in general into two main categories, namely, things perceived by the senses (ὁρατά) and those perceived by the intellect (νοητά). Each category is subdivided into two smaller divisions according to the degree of truth and clarity they exhibit. Geometry and the related sciences (which include astronomy, as is made clear later) are placed in the lower division of the νοητά, because these sciences make use of hypotheses which are assumed and not explained, such as the concepts of odd and even, line and angle, and the various geometrical shapes (510c–d). Moreover they employ figures and diagrams to represent these, so that they may be regarded as the counterparts in the intellectual category to that subdivision of the objects of sense perception which includes such things as shadows and reflections (of the physical objects which constitute the other division of this category), and represents the lowest grade in respect of clarity and truth (510d–e). Thus the triangles, squares, and circles that the mathematician uses in his deductive reasoning from certain basic assumptions are really only imperfect images of the ideal forms of these figures, which can only be apprehended by the mind working through the method of dialectic (διαλεκτική). Such forms are superior because 'the dialectician treats his hypothesis as purely provisional, testing, revising, rejecting, and reconstructing, and gradually ascending step by step to the first principle of all (τὴν τοῦ παντὸς ἀρχήν), without employing any sensible objects to illustrate his reasoning.[128]

This is not the place to enter into a detailed discussion of the famous theory of Ideas (or Forms – both Plato and Aristotle use ἰδέαι or εἴδη indifferently in this sense) which is the keystone of Platonic philosophy; the subject has generated an enormous literature,[129] and remains and doubtless always will remain a matter of keen controversy.[130] Suffice it to say that although the theory can be made to seem plausible for certain classes of physical objects and abstracts such as moral and aesthetic qualities or mathematical relationships (e.g. equality, size, oddness, and the like), difficulties arise in the case of other mathematical and geometrical concepts, and even more in astronomy. Apart from such obvious questions as 'Are there Ideas of all numbers? What is the relation between such Ideas and aggregates of physical objects or the numbers manipulated by a mathematician? How do the ideal point, line, triangle, circle, etc. stand in relation to their hypothetical counterparts in geometrical theory (where absolute truth is attainable – cf. Rep. vii, 527b), and these again to the imperfect realizations of them in the sensible world?' – apart from this, there is the problem of why Plato relegated geometry and the other sciences to the *lower* division of νοητά in the Divided Line, even though the theorems of geometry at least are regarded as absolutely true. If the Ideas, which belong to the upper division of the Line, are the only entities conceivable apart from the things of the phenomenal world (about which no true knowledge is attainable – 529b), there might seem to be no reason for any such division.

Aristotle's answer is that Plato tacitly assumed three classes of entities, the Ideas, the objects of the world of the senses, and 'mathematical intermediates' (τὰ μαθηματικὰ μεταξύ) which comprise the objects of mathematics, and like the Ideas are perfect and unchanging, but are conceived as separate from them.[131] It is true that nowhere in Plato's writings is such a classification explicitly stated, but several passages, as well as the Divided Line analogy, might seem to point in this direction.[132] No attempt can be made here to give the detailed arguments for and against the supposition of 'intermediates', but note must be taken of the implications of this interpretation of Plato's thought for his astronomy. As Aristotle points out (*Met.* B 2, 997*b*12*f.*), if the concept is applied to the objects of astronomical study, then one must postulate another sun and moon and other stars as the perfect counterparts of the visible (and therefore *a priori* imperfect) ones we know; and these can be neither moving (because they belong to the realm of unchanging being) nor unmoving (or else they would not be the objects of astronomy); but this is both inherently and logically absurd.

The only passage that might seem to support Aristotle's imputation of astronomical 'intermediates' to Plato is *Rep.* 529*c*–*d* (but on this see below). Nowhere else in the Platonic dialogues is there any hint of an 'ideal' sun and moon and 'ideal' stars; on the contrary, as we shall see, each of the celestial objects is regarded as a divinity and endowed with a soul. Aristotle's criticism is obviously justified if Plato truly meant to postulate astronomical 'intermediates'. However much it may be argued that he never really came to grips with the problem of fitting astronomy into his ontological scheme,[133] so that his treatment of it is in some respects unsatisfactory, yet it is difficult to believe that he would have chosen to take up a position which laid him open to such criticism.

What is quite certain is Plato's conviction of the importance of mathematical studies in training the soul to look beyond the transitory things of this world to the true reality which can be comprehended by thought alone; such studies are ἀγωγά . . . ἐπὶ τὴν τοῦ ὄντος θέαν . . . πρὸς ἀλήθειαν (525*a*–*b*). Their usefulness in the business of everyday life is not denied, but Plato insists that their real value lies, not in their practical applications, but in their power to quicken the intellect and condition the soul in its quest for truth (525*d*–526*c*). When Glaucon applauds the choice of geometry as the second subject (after arithmetic) in the ideal syllabus, on the grounds of its usefulness to the general in the marshalling of armies and the disposition of camps, he is gently rebuked by Socrates, who points out that a mere smattering of geometrical knowledge would suffice for this (526*d*–*e*). The real reason for studying it ought to be because of the eternal truths it embodies, and because it draws the soul upwards and is productive of a philosophical turn of mind (ὁλκὸν ἄρα, ὦ γενναῖε, ψυχῆς πρὸς ἀλήθειαν εἴη ἂν καὶ ἀπεργαστικὸν φιλοσόφου διανοίας, 527b). Thus it is the *reason* for Glaucon's recommendation of geometry, and not its utility that Socrates criticizes; in fact, he himself endorses its practical uses, but only as 'πάρεργα' (527c).

Next in order after geometry, Socrates at first proposes to put astronomy, and this likewise is welcomed by Glaucon because of its practical utility in conveying information about the seasons and months and years to the farmer, the sailor, and the general (527d). Once again Socrates corrects him, more sharply this time, for being so intent on placating popular opinion by stressing the utility of the subject that he forgets that the whole purpose ('certainly no mean one, but difficult to believe') is to purify a particular organ of the soul which is more valuable than a myriad eyes, because by it alone can truth be perceived (ὄργανόν τι ψυχῆς . . . κρεῖττον ὂν σωθῆναι μυρίων ὀμμάτων· μόνῳ γὰρ αὐτῷ ἀλήθεια ὁρᾶται, 527d–e). Thus Socrates' criticism is again directed against the *motive* for Glaucon's support of astronomy, not against the practice of it nor against observation as such.

The order in which the mathematical sciences are to be taken is now corrected, and stereometry is inserted before astronomy; this, explains Socrates, is better suited (528a–b) to their logical order of development, since the study of stationary solid bodies should come after that of plane figures (geometry) and before that of solid bodies in motion (i.e. astronomy). Hence the latter will stand fourth in order. Glaucon again approves, and, to try to show that he has benefited from Socrates' earlier rebuke, makes a seemingly innocent remark to the effect that this science certainly 'compels the soul to look upwards and leads it away from (terrestrial) matters to that other region' (529a). This, however, will not do for Socrates, who asserts that those who profess to lead men to philosophy treat astronomy in such a way as to make the soul look down (κάτω βλέπειν). When the luckless Glaucon enquires what he means, he is accused of giving the impression of believing that simply gaping up at a painted ceiling will engender real knowledge,[134] whereas, Socrates insists, no learning (μάθημα) can 'make the soul look upwards' except that which is concerned with 'Being and the Unseen' (τὸ ὂν καὶ τὸ ἀόρατον). It makes no difference whether a man gazes up with mouth agape or down with pursed lips,[135] for if he is occupied with the things of the senses (τὰ αἰσθητά) he can never gain true knowledge (ἐπιστήμην γὰρ οὐδὲν ἔχειν τῶν τοιούτων, 529b ad fin.). Glaucon meekly acknowledges his fault and asks Socrates to explain how he thinks astronomy should be studied; it is Socrates' answer to this question which is largely responsible for the grossly distorted view of Plato's attitude to astronomy (or even to all science) held by many modern writers,[136] and which has caused even staunch Platonists considerable embarrassment[137] – entirely unnecessarily.

It is necessary to quote in detail here because it is on these few sections of the *Republic* out of the whole Platonic corpus that the case rests for Plato as a retarding force in the development of Greek astronomy and an opponent of observational science. Socrates says,

> These patterns in the heaven, since they have been embroidered in the visible region, we should indeed consider as the most beautiful and most precise of such phenomena, but as falling far short of the true realities with regard to the revolutions in which the real speed and the real slowness, in their true measure and in all their true forms, are both carried in relation to each other and carry the objects they contain.[138]

These true revolutions, Socrates continues, are only to be grasped by the mind and not by the sight; in fact, one ought to use the variegated heavens as 'paradigms' of them in the same way as one might use the beautifully executed diagrams of a master craftsman to help one to grasp the truth of a geometrical theorem. One would not, however, suppose that the diagrams themselves, for all their perfection, are sufficient to reveal by mere inspection the fundamental mathematical concepts, such as equality and ratio, which underlie them (529d–e). The true astronomer will take a similar view (both literally and figuratively) of the revolutions of the stars. He will grant that they have been created as near perfect

as visible things can be, but he will *not* thereby assume that the
ratios of night to day, of day to month, of month to year, and of
the periods of the other stars (i.e. the planets) to these and to each
other are exactly what they seem to be from observation of the
celestial bodies, and that there is no difference at all between the
real ratios and the apparent ones, so that he is justified in con-
centrating all his attention on the latter (530a–b). 'In fact,' says
Socrates, with deliberately dramatic exaggeration, 'we shall pursue
astronomy just as we do geometry by making use of problems,
and we shall leave the phenomena of the heavens alone,[139] if we
intend, by getting a real grasp of astronomy, to improve the
usefulness of the intellectual faculty inherent in the soul' (530b–c).
This famous remark by Socrates, wrongly interpreted out of its
context, has proved a godsend to the detractors of Plato, who have
pounced on it with glee as proof positive for their case. Glaucon
then comments that the study of astronomy along these lines will
be much harder than it is now, and Socrates agrees but affirms the
necessity for this type of approach in other subjects as well 'if we
are to be of any service as lawgivers' (530c).

It is important to keep in mind the underlying purpose of this
part of the *Republic*, which is to discover the subjects best suited
to form part of the ideal education for the guardians of the
perfect state *and* to discuss the best methods of treating them. The
whole tenor of Plato's exposition is 'propaedeutic'; the emphasis
is on training the soul not to be content with mere knowledge of
the transient phenomena of the visible world (δόξα), but to strive
towards knowledge of the eternal verities that lie behind these
changing appearances. This is the true knowledge (ἐπιστήμη – *cf.*
533e–534a) that ideally the guardians ought to possess. To achieve
it, Plato was convinced of the signal importance of mathematics
as a discipline that compels the mind to deal with abstract concepts
and relations. Mathematics can, of course, in many cases be applied
to concrete things, but its truth does not depend on such applications and can only be grasped in its generality by the exercise of
pure thought. Thus mathematics teaches the soul to look behind
and beyond appearances, and so provides excellent training for
the effort to comprehend true reality and the ideal good, but it is

not *by itself* capable of realizing this final aim, which is reserved for the supreme science of dialectic (see above). Both the Divided Line and the analogy of the Cave serve to illustrate Plato's point in their different ways, the former by emphasizing the complete separation of the world of the senses from the world of thought, and by placing the mathematical sciences in their proper position relative to other forms of knowledge,[140] and the latter by vividly contrasting the shadowy world of appearances with the brightness of true reality.

The importance attached by Plato to mathematics as a pro- paedeutic discipline colours his whole conception of the study of astronomy. He wanted astronomers to pay more attention to the mathematical side of their subject, to go beyond mere observation of celestial phenomena, and to investigate the mathematical relations subsisting between the various orbits of the visible celestial bodies. Convinced that the whole universe operated according to mathematical laws (the cosmology of the *Timaeus* makes this abundantly clear) which could be understood only by the properly trained intelligence, and equally convinced that no certain knowledge could result from the things of the visible world of the senses alone, he naturally (and rightly – see below) stressed the theoretical side of astronomy.[141] Socrates' remark about attacking astronomy by means of problems and leaving the heavens alone is a striking hyperbole,[142] designed to give vivid expression to what was uppermost in Plato's thought, the need to emphasize the mathematical background to astronomy. It argues an astonishing naivety or a wilful blindness to Plato's real purpose when commentators[143] interpret Socrates' words here as proof of Plato's baneful influence on scientific progress.

Taken in its proper context, with due attention paid (as it seldom is) to the last part of the sentence ('if we intend . . . to improve the usefulness of the intellectual faculty inherent in the soul'), Socrates' remark affords not the smallest justification for supposing that he intended to denigrate the role of observation in astronomy. As we have seen, he was perfectly ready to approve the practical utility of all the mathematical sciences (the *Laws* emphasizes this in the case of astronomy – see below), only insisting that this was not the most important reason for including them in the educational syllabus. Adam's opinion that Plato conceived of 'a mathematical οὐρανός of which the visible heavens are but a blurred and imperfect expression in time and space' (*op. cit.*, p. 128), and his attribution to him of a 'magnificent dream of a starry firmament more beautiful and perfect than the visible sky' (App. II, p. 168) reveal a misunderstanding of Plato's point of view. Certainly he wanted to penetrate behind the visible phenomena to the mathematical realities underlying them, but there is no evidence that he believed in astronomical 'intermediates', i.e. 'intelligible realities occupying an intermediate position between sensibles and Ideas' (*ibid.*, p. 167), which, as Aristotle points out (see above), would have been an absurd position for him to take.

There is a sound historical reason, not hitherto appreciated, for Plato's insistence on the importance of the mathematical side of astronomical studies. As we have seen, there had been accumulating during the sixth and fifth centuries BC a mass of necessarily crude, but nonetheless practically useful, observational material, which resulted in the appearance of the first astronomical calendars ('parapegmata' – see pp. 84ff.) and enabled Euctemon and Meton to fix reasonably accurate values for the solar year and the astronomical seasons. There is a limit, however, to the usefulness of continuing to collect such empirical data, unless it is accompanied by a corresponding advance in understanding its significance and fitting it into a general astronomical scheme by applying mathematical techniques to its evaluation. There is good reason to believe that the end of the fifth and the beginning of the fourth century BC saw this limit reached in the development of Greek astronomy. What was wanted now was for astronomers to sit down and do some hard thinking about astronomical theory, so as to make the best use of the observations they already had, and evolve a mathematically based system which would take into account not only the long-known variations in the aspects of the night sky, but also the irregularities apparent in the movements of sun, moon, and the recently discovered planets. The concept of the spherical earth in the middle of the celestial sphere obviously provided a suitable framework, but much more theoretical work was needed to correlate the observational material into it.

This is precisely what Plato realized, and this is why he urged the astronomers to concentrate on the mathematical side of their subject and study the real mathematical relations lying behind the visible phenomena, to the extent even of temporarily calling a halt to the mere accumulation of more observations.[144] He can hardly have failed to be aware of Eudoxus' efforts in this direction (see below), and it seems certain that he recognized that they were an improvement on the Pythagorean notion of a planetary earth orbiting round the central fire.[145] His stated conception of the 'logos' of astronomy as being concerned with the relative speeds of the celestial bodies (*Gorgias* 451c) would agree well with the Eudoxan attempt to construct a mathematical scheme for the planetary movements, and is consistent with the story reported by Simplicius (*in De Caelo* ii, 12, 219a23) on the authority of Sosigenes, who in turn took it from Eudemus, that Plato set it as a problem to those interested in such matters to determine what uniform and ordered movements should be assumed if the observed motions of the planets were to be accounted for. The fruits of Plato's eminently sensible recommendation on how astronomy should be studied can be seen in the comparatively rapid development of mathematical astronomy after him, culminating in the definitive system first sketched by Hipparchus some two hundred years after Plato's death, and not finally completed until the work of Ptolemy about three hundred years later. To denigrate Plato as a retarding influence on astronomical science is not only to misinterpret his attitude entirely, but also to ignore the actual history of this development.[146]

To return to the *Republic*: Socrates goes on to discuss the science of harmonics or musical theory, and points out that harmonics is to the ear what astronomy is to the eye – they are sister sciences 'as the Pythagoreans say and as we too agree' (530d); the same advice is given to the practitioners of this science as to the astronomers, namely, to concentrate more on the theoretical side of their subject and not expend endless trouble over physical measurements alone to the neglect of theoretical problems (531a–c).[147]

Only thus will the two sciences justify their inclusion in the educational syllabus, the whole point of which is to give the mind the best training for grasping the intricacies of the dialectical method, which Socrates then proceeds to describe (531d–535).

There is only one other passage in the *Republic* that has any astronomical significance, and this occurs in the myth of Er in the tenth book. Here Socrates tells the story of Er, son of Armenius, whose soul accompanied the other souls of the dead to their final judgment, but then miraculously returned to his body. It had been picked up on a battlefield and after twelve days was about to be burnt on a pyre, when suddenly he came back to life and related his experiences 'after death'. The mythical tone of the whole narrative is set right at the beginning (614c), when Er describes how the souls came 'to a certain magic place' (εἰς τόπον τινὰ δαιμόνιον) where there were two chasms in the earth facing two chasms in the sky, through which twin processions of souls trooped continuously, the just being directed upwards to receive their rewards, and the unjust downwards for their punishments. Both groups (except for some particularly wicked sinners) after the proper span of time returned for their reincarnation in future lives. When the souls had remained seven days at this clearing station, they set out en masse and on the fourth day arrived at a place from which 'they saw from above[148] a straight light, stretched through the whole heaven and earth, something like a pillar, yet very similar to the rainbow, but brighter and clearer' (616b). After another day's march 'they saw there at the middle of the light the extremities of its chains stretched from the heaven – for this light was a fastening of the heaven, like the under-braces of triremes, and so held together the whole revolution – and from the extremities was stretched the spindle of Necessity, by means of which all the revolutions[149] are turned'; the shaft and the hook of the spindle are made of adamant (the hardest metal) and the whorl (σφόνδυλος, the thick circular disc that serves to balance the spindle and allows it to be rotated easily by the fingers) of this and other materials.

Plato then goes on to describe the structure of the whorl, which is 'hollow and scooped out all through' and, in fact, consists of eight whorls fitting closely one inside the other, so that their lips seen from above form one continuous surface, through the centre of which passes the shaft of the spindle (616d–e).[150] The whorls are arranged in order, first by the breadth of their lips, next by their characteristic appearances, and lastly by their manner of revolution (616e–617b). The whole spindle revolves with the same circular motion, but within the revolution of the whole the seven inner circles rotate with varying speeds in the opposite direction to the movement of the whole. Finally, as further poetic embellishments, we are told that the spindle is turned on the knees of Necessity, that on each of the whorls is perched a Siren uttering one note and all together constituting one harmony, and that the three daughters of Necessity, the Fates (Clotho, Atropos, and Lachesis), sit round about and keep the spindle and its eight whorls turning with their hands (617b–d).

This highly fanciful, visionary picture has given rise to numerous equally fanciful interpretations at the hands of commentators both ancient and modern, and desperate attempts have been made to find some sort of scientific coherence in Er's description. The difficulties, however, are insuperable. Plato never makes it clear (no doubt intentionally) where he supposes the souls to be located and from what standpoint Necessity's spindle is viewed, whether from outside the universe altogether, from the sphere of the fixed stars, from the 'real', outer surface of the earth, as in the *Phaedo* (see above), from the centre of the universe, or from the north pole. All these have been suggested. The 'straight light' (variously interpreted as the Milky Way or a shaft of light enclosing the north-south axis of the universe[151]) is in one breath described as 'stretched through the whole heaven and earth'[152] and in the next as binding the whole revolution of the heaven 'like the under-braces of triremes'.[153] Similarly, the spindle is depicted as 'extended from the extremities' of the light (ἐκ δὲ τῶν ἄκρων τεταμένον), and yet it is turned on the knees of Necessity.

There would seem to be at least three separate images the details of which are kaleidoscoped in Plato's mind: first, there is the pillar of light which is somehow omnipresent in the universe; second, there is the image of the universe as a construction like a ship's hull which needs to be held together (this may have been suggested by the great circles of the celestial sphere, such as the zodiacal band, the equator and the meridians, which, as it were, bind together the whole); and third, there is the spindle with its composite whorl turned by the spinning Fates – a combination of traditional mythology and a typically Platonic use of simile drawn from human craft occupations. It is impossible to disentangle these images so as to form one coherent, scientifically valid picture; this is the celestial mechanics of myth, as Plato makes clear by introducing the Fates to provide the motive power for the whole, not to mention the singing Sirens on each whorl. The latter are clearly an extension of the Pythagorean 'harmony of the spheres' (see above), just as the goddess Necessity recalls Parmenides' 'Necessity holding the limits of the stars' (*DK* 28 B10) and his 'divinity in the middle which controls everything' (*DK* 28 B12), and also Empedocles' goddess 'Love (Φιλότης) in the middle of the eddy' (*DK* 31 B35).

Pythagorean influence is likewise shown in the one astronomically significant part of the description, that of the eight whorls. It is clear that these are meant to represent schematically the orbits of the moon, sun, the five planets (although these are not mentioned by name here; Venus and Mercury are mentioned in *Tim.* 38d, and all of them in *Epin.* 987b–c – see below), and the fixed stars. Not only is the outermost whorl described as 'spangled' (ποικίλος) and as having the broadest rim (obviously to accommodate the myriad stars which it includes), but the eighth (from the outside) is said to receive its colour from the light of the seventh (certainly Anaxagoras and probably Parmenides already knew that the moon shines by reflected sunlight – see above), and the fourth (i.e. Mars) is characterized as 'reddish' (ὑπέρυθρον). Moreover, the whole spindle is depicted as turning in a circle 'with the same motion' (τὴν αὐτὴν φοράν) at the greatest speed (this foreshadows the circle of the Same in the *Timaeus*, producing the diurnal rotation of the whole heavens), while the seven inner whorls 'revolve slowly in the opposite direction to the whole' (this foreshadows the circle of the Different in the *Timaeus*, which carries the orbits of sun, moon, and planets—see below).

As regards the speeds of revolution of the whorls, the eighth (the moon) is stated to revolve the fastest (of the seven inner ones), then the seventh, sixth, and fifth together at the same speed (that is the sun, Mercury, and Venus, which in *Tim.* 38*d* are again given the same angular velocity – see below), then the fourth (Mars), then the third (Jupiter), and then the second (Saturn). Mars is further characterized as 'revolving, as it seemed to them (the souls), ἐπανακυκλούμενον' (617*b*). The last word is usually translated 'with a counter-revolution',[154] meaning counter to the revolution of the outer whorl representing the sphere of the fixed stars. However, in the previous sentence it has already been stated that the seven inner whorls revolve in the opposite direction to the revolution of the whole. Why, as Cornford asks, should it be mentioned again, and why should it be connected particularly with Mars? Cornford's own explanation is not at all convincing (see below, pp. 124*f.*) and necessitates taking ἐπανακυκλούμενον as qualifying both the third and second whorls as well as the fourth, in defiance of the text which clearly confines the word to the fourth alone.

What phenomenon connected with its apparent motion might cause Plato to distinguish Mars from the remaining two superior planets by describing it as ἐπανακυκλούμενον? There is one possible answer; whereas Jupiter and Saturn make one loop in the sky every year during the 29 and nearly 12 years respectively that they take for a complete circuit of the zodiac, Mars, taking just less than 2 years, makes only one loop during this period and makes more than one circuit of the heavens between one loop and the next – so that, in contrast to the other two, it could in a sense be described as making (part of) an *additional* circle. I suggest, then, that ἐπανακυκλούμενον should be translated as 'making an additional revolution', and that ἐπι- should be taken in the same sense as in ἐπινομοθετέω (*Laws* 779*d*, 'enact in addition'), ἐπιδίδωμι (*Laws* 944*a*, 'give in addition', i.e. as a dowry), ἐπίτριτος and ἐπόγδοος (*Tim.* 36*a*, 'one plus a third', 'one plus an eighth', i.e. in the ratios 4:3 and 9:8), ἐπίκτητος (*Laws* 924*a*, 'additional' property), etc., etc. The verb ἐπανακυκλέω seems to occur only here in classical Greek, but the noun ἐπανακυκλήσεις occurs at *Tim.* 40*c* where (as we shall see) it can certainly bear the same meaning; Galen (ed. Kuhn, vol. vii, p. 412) uses the verb of a recurring fever. Thus Plato singles out the whorl representing Mars for special mention not only because it revolves some six times faster than that representing Jupiter and nearly sixteen times faster than that of Saturn (corresponding to the sidereal periods of these planets), but also because of the different pattern of its loops. This implies that he was familiar with the phenomenon of planetary retrogradation, and in fact there is no good reason to doubt this (see below, on the *Timaeus*); Eudoxus' system of concentric spheres was expressly designed to account for retrogradations and it is hardly conceivable that Plato was totally ignorant of it – oddly enough, Mars is the planet for which the system gives the worst results as far as we are able to reconstruct it.

There is no hint in the *Republic* of the fact that the zodiac is inclined to the plane of the equator. This may simply be because it would not have suited Plato's picture of the spindle to introduce this concept here, whereas in the *Timaeus* it forms an integral part of the description of the celestial sphere; but it may also be the case that he had not yet fully grasped the idea, since on the normal dating of his works there is at least twenty-five years between the *Republic* and the *Timaeus*. The arrangement of the whorls all in one plane with a common centre bears a strong resemblance to the Philolaic system, where (as we have seen) according to the best authorities there is also no mention of inclined orbits. The important difference is that Plato has dropped the notion of the planetary earth and counter-earth as also the hypothetical central fire, and made the earth the centre of the universe (this is not stated in so many words, but no other interpretation will fit the description). The order of the whorls, fixed stars, five planets, sun, and moon, remains the same as in the Philolaic system. Both the latter and Plato's vision are highly schematic; the different breadths of the lips of the whorls may be intended to symbolize the different distances of the planetary orbits from the earth or their different periods of revolution, but there is no discernible correlation between the order assigned to the breadths and any astronomical facts. Similarly, in the Philolaic system we have no evidence of the periods assigned to the planets (see above). Both systems appear to be based largely on *a priori* conceptions, but historically the Platonic picture in the *Republic* reveals itself as a theoretical improvement on the Philolaic scheme and as a step towards the full understanding of the concept of the celestial sphere.

In the *Phaedrus* too there is a passage that has some relevance to astronomy, again heavily disguised in mythical language. As part of the comparison of the soul to a chariot with a driver and two horses (246a), we are given a picture of Zeus, the ruler and over-seer of everything, in his winged chariot leading the whole host of gods and daemons ranged behind him in eleven groups, 'for Hestia alone remains behind in the house of the gods' (μένει γὰρ 'Εστία ἐν θεῶν οἴκῳ μόνη), while each of the other twelve Olympian deities leads its own group 'according to the order ordained for each' (247a). 'Within the heavens are many marvellous sights and highways' (πολλαὶ . . . διέξοδοι ἐντὸς οὐρανοῦ) used by the gods to reach 'the back of the heavens' (ἐπὶ τῷ τοῦ οὐρανοῦ νώτῳ, 247b – the last is the word also used in the description of the spindle of Necessity to depict the solid appearance of the lips of the whorls as seen from above, νῶτον συνεχές, *Rep.* x, 616e). The gods make this journey easily in their well-trained equipages, but the souls of mortals have the utmost difficulty in accomplishing it because of the unruliness of their steeds. Once arrived at their destination, the immortal souls stand and are carried round by the revolution (αὐτὰς περιάγει ἡ περιφορά) and gaze on the things outside the heaven (247b–c).

The ancient commentators are unanimous in taking this as a symbolical representation of the arrangement of the universe: Zeus represents the sphere of the fixed stars, Hestia the earth, and ten other gods respectively the five planets, sun, moon, aether (*cf. Epin.* 984b), air, and water. This is not very convincing,[155] and Hackforth may well be right in discounting any specific astrono-mical interpretation here and seeing merely a reference to the altar of the twelve gods mentioned by Thucydides (vi, 54) as being set up by the younger Pisistratus in the agora.[156] On the other hand, the emphasis on the number twelve (unnecessary for the general picture) may have an astronomical significance in that it may refer to the twelve signs of the zodiac and twelve deities connected with them;[157] there is no doubt that the concept of the zodiac was becoming known in Plato's time, and Eudoxus was familiar with it (see below). At any rate it is clear that Hestia must be envisaged as being left behind in the centrally placed earth (presumably called the 'house of the gods' because on it were located the gods' temples, not to mention Mt Olympus).[158] No other interpretation will fit the picture of the gods' ascending easily to the uttermost extremity of the vault of heaven (ἄκραν ἐπὶ τὴν ὑπουράνιον ἀψῖδα πορεύονται πρὸς ἄναντες, 247a–b), while the refractory horse of a mortal soul's chariot inclines downwards towards the earth (ἐπὶ τὴν γῆν ῥέπων, 247b). In particular, it is impossible that Hestia here should be taken as referring to the Pythagorean central fire (also sometimes called Hestia, see above), as Boeckh thought (Kleine Schriften, vol. iii, p. 289); not only is a planetary earth and counter-earth completely foreign to Plato's astronomical views, but it would be singularly inapposite in this context, since each of these bodies would have to be envisaged as being the leaders of groups of souls.

In the Politicus myth (269a–247d), we have the intriguing picture of the whole universe rotating now in one direction and now in the opposite, according to whether the god is in direct charge of it or has temporarily left it to its own devices, so that the risings and settings of sun, moon, and stars occur at diametrically opposite places in the two cycles (conceived of as lasting for myriads of periods, 270a). All this adds little to our knowledge of Plato's astronomy; it is a highly fanciful and in places intentionally amusing account of a time when the mortal life-cycle was reversed and hoary old men yearly became darker-haired, more vigorous and younger until they eventually disappeared as tiny infants (270e)! It serves to confirm that for Plato the basic revolution of the universe (i.e. that of the sphere of the fixed stars, the circle of the Same in the Timaeus – see below) is the underlying cause of all visible phenomena, and it may provide evidence of an increased awareness on his part of the importance of astronomical cycles, particularly that of the sun.

In 270*b–c* it is stated that 'This reversal (i.e. of the universal rotation) ought to be considered the greatest and most complete turning of all the turnings that take place round the heavens', ταύτην τὴν μεταβολὴν ἡγεῖσθαι δεῖ τῶν περὶ τὸν οὐρανὸν γιγνομένων τροπῶν πασῶν εἶναι μεγίστην καὶ τελεωτάτην τροπήν. Although the word 'turning' (τροπή) in an astronomical context can be used in a general sense of the courses of the moon and planets as well (*cf. Tim.* 39*d*; Arist., *Meteor.* 353*b*8), it can hardly have failed to remind his listeners of the 'turnings of the sun' (τροπαὶ ἡλίου – see p. 32), i.e. the solstices, with which, of course, Plato himself was familiar (*cf. Laws* xii, 945*e*, μετὰ τροπὰς ἡλίου τὰς ἐκ θέρους εἰς χειμῶνα). It is probably an intentional pun; a similar play on words occurs at 271*c*, this time with the sun actually mentioned (τὴν μὲν γὰρ τῶν ἄστρων καὶ ἡλίου μεταβολὴν δῆλον ὡς ἐν ἑκατέραις συμπίπτει ταῖς τροπαῖς γίγνεσθαι). Possibly his growing interest in astronomy and the contemporary concern to rectify the calendar and correlate lunar and solar cycles may have prompted his choice of words.

We come now to the dialogue in which Plato comes nearest to giving us a detailed description of his astronomical opinions, the *Timaeus*, which, after some preliminary conversation between Socrates, Timaeus, Hermocrates and Critias (including a brief account of the story of Atlantis as told to Solon by the Egyptian priests), consists of an unbroken monologue by Timaeus on the subject of the creation of the universe and all its contents by the divine craftsman or Demiurge (ὁ δημιουργός). There is no evidence that Timaeus of Locri in Italy (*cf.* 20*a*) was a historical personage,[159] and it seems certain that he was merely a convenient mouthpiece for Plato's own views; the fact that he is described by Critias as the best astronomer in the company (27*a*) no doubt reflects the increasing importance that Plato attached to astronomical studies in his later years (see above).

There has been considerable discussion as to how far Plato intended his account of the constitution of the universe to be taken literally, since in certain respects (e.g. the generation of the universe, the concept of time, and the self-motion of the soul) the doctrines of the *Timaeus* are not only apparently contradictory in themselves, but also incompatible with Plato's thought as

propounded in other dialogues.[160] Without entering into the
philosophical implications, one can safely say that as regards astronomy it is highly improbable that every detail of the Demiurge's
creative procedures is meant to be interpreted as Plato's literal
belief, but that there is no reason to doubt that the resultant
picture as a whole, based on the concept of the celestial sphere,
corresponds to his own views of the structure of the universe, and
that these show a marked development compared with those
ascertainable from earlier dialogues.

Thus it is obvious that the picture of the Demiurge fashioning
the universe out of the four basic elements (fire, earth, water, and
air) in geometrical proportion (32a–c) and making it a smoothly
spherical being that possesses no bodily organs of any description
(33b–d) is purely symbolical, owing much to Parmenides and
Empedocles (see above). Likewise, the description of the composition of the stuff out of which the World-Soul is made, its
division into two combined geometrical proportions (1, 2, 4, 8
and 1, 3, 9, 27), and the insertion of more parts of the original
mixture in harmonic and arithmetic mean between the numbers
of the original series (35a–36b) – all this[161] serves to symbolize
Plato's conviction (shared by the Pythagoreans) that mathematical
harmony and proportion lie at the root of all *real* knowledge about
the universe. Similarly, it would be naive to suppose that he meant
his account of the Demiurge's taking a strip of the soul-stuff,
bending it round in two circles, joining them like the letter 'chi',
and then splitting one of them into seven parts (36b–d) to be taken
literally; again this is a symbolic representation of a fundamental
feature of the celestial sphere, i.e. the outer surface representing
the fixed stars and, inclined at an angle to its equator, the zodiacal
belt within which move the five planets, sun, and moon. The
description of how this arrangement might have come about is
well suited to a mythical story. Timaeus himself is at pains to
point out that, in dealing with such a vast subject, we as mere
mortals cannot expect to attain to absolute truth, but must be
content with at best a 'likely story' (τὸν εἰκότα μῦθον, 29d[162]).

To turn now to the details of the account in so far as they are
relevant to astronomy: the whole heaven or universe or cosmos

(Plato uses the terms οὐρανός and κόσμος as synonyms, 28b) is a single, unique creation (31b – in opposition to the Atomist conception of a plurality of possible worlds), turned by its creator in the shape of a sphere (σφαιροειδές . . . κυκλοτερὲς αὐτὸ ἐτορνεύσατο, 33b), and having a circular, revolutionary motion assigned to it (34a). In the middle of this 'one, solitary, circular heaven turning in a circle' the god established 'soul' (ψυχή) which he 'extended throughout the whole and with which he also covered the body (of the universe) round on the outside'; and 'he created it a blessed god' (34b). The idea of the sphericity of the universe was not new and certainly goes back to Parmenides (see p. 51). Similarly, the notion of soul as a controlling force (cf. 34c – soul as the 'master and ruler of body') set in the middle and extending to the outermost circumference is an obvious extension of the Parmenidean 'Necessity holding the limits of the stars' (cf. p. 111), and also shows affinities with the Pythagorean conception of the controlling force situated in the central fire; but Plato's World-Soul is a more sophisticated concept.

There follows the passage already mentioned describing the composition of the World-Soul and the mathematical proportions in which the Demiurge handles the mixture (35a–36b), which at the end of his manipulations is entirely used up (οὕτως ἤδη πᾶν κατανηλώκει). The resultant compound (Plato uses the general word σύστασις, but apparently had in mind a long, wide band of soul-stuff) is now split lengthways into two, the two strips are bent round in circles, their respective ends being joined up, and made to lie across each other in the shape of the letter 'chi', being fastened to each other at the intersection points on the plane passing through their centres (36b–c). Both circles are comprehended 'in the motion that is carried round uniformly in the same place' (36c), and the Demiurge gave to the outer one the name and nature of the Same and caused it to revolve 'by way of the side towards the right' (κατὰ πλευρὰν ἐπὶ δεξιά), and to the inner one the name and nature of the Different and caused it to revolve 'by way of the diagonal towards the left' (κατὰ διάμετρον ἐπ'ἀριστερά). 'But he gave the superior power to the revolution of the Same and uniform' (κράτος δ'ἔδωκεν τῇ ταὐτοῦ καὶ ὁμοίου περιφορᾷ), for this he left undivided, whereas the inner one he split 'into seven unequal circles according to the several intervals of the double and triple proportion'. This refers to the combined geometrical progressions, already mentioned above, forming a series 1, 2, 3, 4, 8, 9, 27. 'And he laid it down that they should move in circles in opposite senses to each other (κατὰ τἀναντία μὲν ἀλλήλοις), but three should be similar in speed, while the other four should be dissimilar to one another and to the three, though moving in due proportion' (36d).

Now it is clear that the basic image that lies behind this description is that of the celestial sphere with two of its main circles, the equatorial circle and the zodiac, intersecting each other at a fixed angle (equal to the obliquity of the ecliptic) at two points diametrically opposite the common centre, to give the shape of the letter 'chi' lying on its side (see Fig. 4). The equatorial circle is the circle of the Same, which represents the diurnal revolution of the entire celestial sphere and is superimposed on the individual movements of all the heavenly bodies, while the zodiac corresponds to the circle of the Different, which is split into seven circles, each having its own different motion, but all moving in a general sense in the opposite direction to the revolution of the Same. It should, however, be equally clear that the description of the Demiurge's procedure is largely symbolical, the details cannot be taken literally, and it is highly improbable that Plato had any other physical model at the back of his mind except a simple celestial globe.

For example, we are told that the Demiurge took the strip of circular soul-stuff that constituted the Different and split it into seven *unequal* circles. This, of course, is a physical impossibility unless the material is somehow transferred, compressed or extended into the different circles, and even so, if they are to be arranged in an order corresponding to the series 1, 2, 3, 4, 8, 9, 27, what becomes of the original strip? Obviously, the whole picture is a symbolical attempt to reconcile the two distinct and difficult concepts of the zodiac as a kind of composite great circle of the celestial sphere within which the sun, moon, and the five planets are seen to carry out their revolutions relative to the fixed stars, and the actual orbits of the bodies envisaged as arranged at different spatial depths between the central earth and the outermost sphere of the fixed stars.

Cornford in particular is obsessed with the idea that Plato is describing an actual physical model,[163] and a very sophisticated one at that, apparently combining the features of an armillary sphere, a planispheric astrolabe, and an orrery.[164] Now instruments *were* constructed (known as planetaria[165]) in the eighteenth century that simulated mechanically the movements of sun, moon, and planets as well as showing the chief features of the celestial sphere, thus combining the functions of an armillary sphere and an orrery; and it is possible that the 'sphere' of Archimedes (mentioned several times by Cicero, e.g. *De Rep.* i, 21–2 and *Tusc. Disp.* i, 63) was an early example of this type (*cf.* A. Schlachter, *Der Globus*, 1927, (Stoicheia, Heft 8), pp. 49*f.*). But this was in the last decades of the third century BC, when the celestial sphere was already a well-known concept, and one of the chief problems that faced astronomers was to account for the apparently irregular motions of the planets. It is absurd to suppose that such a relatively complicated instrument would be familiar in Plato's time, when the whole concept of the celestial sphere was still a novelty and very little indeed was known about the planets (*cf.* 39*c*). Cornford has evidently been misled by the tendency of ancient commentators to find in Plato allusions to astronomical instruments and ideas with which they themselves were familiar, but which are certainly anachronistic for his time. Thus Proclus (*in Tim.* iv, 284*b* = p. 145 ed. Diehl) refers to the armillary sphere and the astrolabe[166] in commenting on *Tim.* 40*c–d* (on which, see below). Chalcidius (p. 163 ed. Wrobel = 145 ed. Waszink, 1962), in flat contradiction to the text, supposes that the Demiurge was dealing with *three* circles including a meridian circle (Cornford apparently accepts this, p. 77). Finally Theon (*Ad Leg. Plat. Util.*, pp. 160*ff.* ed. Hiller) imports the whole mechanism of epicycles and eccentrics into Plato's astronomy, a procedure which is rightly rejected by Proclus, who points out that these devices are completely foreign to Plato's astronomical ideas (*op. cit.* iii, 221*f*; iv, 258*e*; *cf.* Heath, *Arist.*, p. 166).

Cornford makes much of the fact that the second ring (that of the Different) is described as being 'inside' (ἐντός) the first (36c), which he says (p. 74) 'is appropriate only to a material model'. This, however, is misconceived. Since there is nothing outside the sphere of the universe (32c; 33c), since the natural motion of the latter is uniform rotation in its own place (34b), and since the circle of the Same is the outermost circle, while the orbits of sun, moon, and planets are situated between this and the central earth (see below – 38c–d), what other designation could be given to the circle of the Different (which embraces also the planetary movements), relative to the Same, than that of the 'inner' one? Plato no more had an actual, physical model of a planetarium before his eyes when he wrote this part of the *Timaeus* than he had a real, composite spindle of the type described in the myth of Er; the most he could have had was a simple celestial globe with some constellations and the main circles of the celestial sphere marked on it (such as was certainly used by Eudoxus – see below), and it was the relative positions of two of these, the equator and the ecliptic, that suggested to him the simile of the letter 'chi'.

The circle of the Same is described as revolving 'by way of the side towards the right' and that of the Different 'by way of the diagonal towards the left' (36c). What is meant is clear from Fig. 10. In the rectangle ABCD formed by the two tropics (AB and DC), the diurnal rotation of the whole sphere (represented by the arrows on AB, DC, and the equator EF) is in the direction of the sides AB and DC), and the annual revolution of the planets along the zodiac is in the opposite direction along the diagonal BD. In terms of compass directions this would mean that east is on the left, but Plato is not here concerned with terrestrial directions (which he is careful not to mention at all); his language is symbolical and the picture presented is schematic only. Since in Pythagorean thought the right is listed in the same column as the one, the male, the straight, and the good, in opposition to the left, plurality, the female, the crooked, and the bad (Arist., *Met.* A 986a24ff.), it was obviously appropriate to have the superior circle of the Same revolving to the right, and the subordinate circle of the Different to the left. Thus there is no contradiction here with

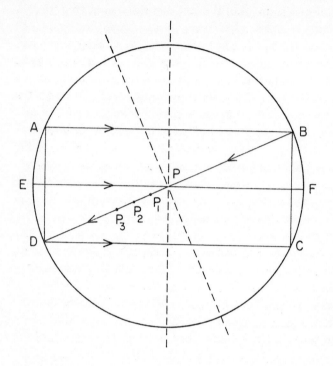

Fig. 10. Diagram illustrating the spiral movement of the planetary bodies referred to at Tim. *39a ad fin.*

the statement in the *Laws* (760*d*) that towards the right means eastwards, or with that in the *Epinomis* (987*b*) to the effect that the proper motions of the planets are towards the right. In both these latter cases the language is that of everyday life, and the viewpoint is that of a terrestrial observer facing north (as in taking omens – the right-hand side, the east, being traditionally regarded as the more auspicious side),[167] not that appropriate to a mythical Demiurge fashioning the universe.

The seven circles into which the Different is split evidently symbolize the seven orbits of the five planets, sun, and moon; some of these (it is not stated which) move 'in opposite senses to each other' (κατὰ τἀναντία μὲν ἀλλήλοις – the significance of this expression will be discussed below in connection with a similar phrase in 38*d*), 'but in speed' (i.e. angular velocity, since all the planets are at different distances from the earth) three are equal (these are the sun, Venus, and Mercury as is made clear in 38*d*), while the remaining four (moon, Mars, Jupiter, and Saturn) are unequal both to each other and the first three (36*d*). Evidently, by this time some knowledge of the different planetary periods was available to the Greek astronomers, and Eudoxus certainly required these data (probably obtained from Babylonian sources) for the working out of his system of concentric spheres (see below).

In 37*d–e*, Timaeus describes how the concept of time developed; 'days and nights and months and years did not exist before the heaven came into being' (πρὶν οὐρανὸν γενέσθαι). Later (38*b*) we are told that 'Time was generated with the heaven' (μετ'οὐρανοῦ). For the Greek thinkers, Time was inseparably connected with periodic motion,[168] and the most convenient method of measuring it was by the cyclic recurrence of uniform circular movement – hence the importance of the celestial bodies (especially sun and moon), the revolutions of which could be counted and thus provide units of time; indeed, the accurate determination of time was probably the chief motive for the development of astronomy as a science. In 38*c*, it is expressly stated that 'sun and moon and five other stars, having the name "wanderers" (πλανητά – or according to some manuscripts πλανῆται or πλάνητες), came into being for the distinguishing and preservation of the numbers of Time'.

The account then continues:

Having made bodies for each of them [i.e. sun, moon, and planets] the god placed these seven bodies in the seven revolutions which the circuit of the Different made, the moon in the first (circle) round the earth, the sun in the second above the earth, and the Morning Star [i.e. Venus] and the star called sacred to Hermes [i.e. Mercury] in circles which in speed move at an equal pace with the sun, but possess the opposite power to it;[169] whence the sun and the star of Hermes and the Morning Star both overtake and by the same token are overtaken by each other (38*c–d*).

The astronomical phenomenon referred to here is the fact that, whereas the outer planets (Mars, Jupiter, and Saturn) can at different times of the year be seen at any angular distance (from 0° to 180°) from the sun, Mercury and Venus are always observed to be close to the sun, appearing as morning or evening stars relatively low down on the eastern or western horizon, and never exceed an angular distance of 28° and 47° respectively from it. The reason for this is that the orbits of these two planets lie wholly within the orbit of the earth in the heliocentric universe (while those of the outer planets lie wholly without), or in other words in the geocentric universe the orbits of Mercury and Venus must be closer to the earth than that of the sun, while those of the outer planets must be further away from it.

How far did Plato really understand this phenomenon? The answer to this question depends on our interpretation of two puzzling phrases in the account of the circles of the Different, first (36d4) that some of the seven circles move 'in opposite senses to each other', and second (38d4) that the circles of Mercury and Venus 'possess the opposite power' to the circle of the sun. If these phrases are taken together in their natural sense, it would seem that Plato is saying that Mercury and Venus revolve in the opposite direction to the sun's revolution (i.e. from east to west, as against the sun's annual motion relative to the fixed stars from west to east), in which case they ought to be seen at all angular distances from the sun without being restricted to the above limits; but this is contrary to the facts of observation, and so Plato's words reveal his deficient knowledge of astronomy.

Both ancient and modern commentators have suggested various interpretations, all more or less far-fetched and none convincing,[170] to try to save Plato's astronomical reputation; of these, Cornford's (pp. 8off.; 106ff.) is undoubtedly the most ingenious and the most carefully worked out. Using the analogy of a moving staircase, bent round to form a continuous circle, with seven passengers on it to represent the general motion (west to east) of the seven circles of the Different, Cornford suggests that these may be divided into two groups. Group (a), comprising the sun, Mercury, and Venus, stands still on the staircase, and is collectively carried down it.

Group (*b*) comprises the moon, Mars, Jupiter, and Saturn, each of which has its own individual motion on the staircase as well as being subject to the general motion downwards. The moon runs quickly *down* the staircase, passing all the others nearly thirteen times in a complete circuit (i.e. in one year), while the other three walk slowly *up* it at different speeds, but are still carried down by the movement of the staircase itself. Only thus, according to Cornford, can it be true to say that the circles move 'in opposite senses relatively to one another' (p. 86). He supports this interpretation by means of the passage describing the revolutions of the whorls of the spindle of Necessity, where it is stated that the fourth whorl (representing Mars) moves 'as it seemed to them (the souls) with a counter-revolution' (617*b*, ὡς σφίσι φαίνεσθαι ἐπανακυκλούμενον τὸν τέταρτον);[171] in this 'counter-revolution' (which he takes to apply also to the third and second whorls, those of Jupiter and Saturn), Cornford sees a reference to the opposing directions of the movements of moon, sun, Venus, and Mercury on the one hand and the three outer planets on the other (p. 88).

The ingenuity of this interpretation should not blind us to the fact that it is based on some large assumptions far from adequately supported by Plato's actual words. To begin with, it entails the assumption of 'a third force which modifies the motion of some of the planets' (p. 87; *cf.* top p. 110), causing, for example, the moon to move quickly in one direction and Mars, Jupiter, and Saturn more slowly in the opposite direction. Yet there is nothing in the text to warrant this, and in fact the whole tenor of the description shows that Plato is thinking of only the two motions of the Same and the Different, the combination of which imparts to each body a spiral twist (see on 39*a*, below). It is true that, as we shall see, Mercury and Venus have an additional peculiarity, but there is no justification for supposing that the disputed phrase 'in opposite senses to each other' refers to the moon as opposed to the outer planets. The single other motion imputed to some of the celestial bodies is rotation on its own axis because each is endowed with a divine soul (see on 40*a*–*b*, below); but this is a completely different concept from the third motion envisaged by Cornford, since it does not affect their translational movement across the sky.

Again, to assume that the singling out of the whorl of Mars for special comment in the description of the spindle of Necessity (which is susceptible of a logical explanation – see above) must also apply to the whorls of Jupiter and Saturn, is a cavalier method of treating the Greek which says no such thing. Moreover, the supposition (implicit in Cornford's interpretation) that Plato distinguished as we do between the outer planets and the inner ones is expressly contradicted by the text; in 38d we are distinctly told that the moon is in the first orbit round the earth, the sun in the second (cf. 39b4), and then follow Venus and Mercury and 'the others', which Plato does not particularize here. This is also the order in which the whorls of the spindle of Necessity are arranged. Thus, as Heath rightly remarks (Arist., p. 168), the modern distinction between superior and inferior planets is not made by Plato at all.

Cornford's explanation of the statement that Venus and Mercury 'possess the contrary power' to the sun is equally open to objections and leads to some confusing contradictions. He rightly sees that this 'contrary power' is evoked to account for the fact that these two planets are never seen far from the sun, first on one side and then on the other, and that this is what Plato means when he says that they and the sun 'overtake and are overtaken by one another' (p. 106). But in that case how can Venus, Mercury, and the sun be regarded as a group standing still on the moving staircase? At some points in the process of overtaking or being overtaken they must move in opposite directions relative to the sun, and this I believe is what Plato had in mind when speaking of their ἐναντίαν δύναμιν. Again, Cornford insists that the 'contrary power' is not confined to Venus and Mercury but is possessed by the outer planets also, and that the latter exercise it continuously (being constantly overtaken by the sun) but the former only intermittently. The awkward fact that the text expressly connects the power with Venus and Mercury only is explained away by saying that Plato did not in fact deny it to the remaining planets, but was not concerned with them for the moment. This highly improbable interpretation is forced on Cornford by his tacit assumption that Plato made the same distinction as we do between the superior and the inferior planets in relation to the sun – but this is a totally false assumption (see above).

Even so, Cornford's account is notably confused as to the nature of this power, its results, and possible causes of these results. He apparently approves of the view of the ancient commentators that the reason for the behaviour of Venus and Mercury is that they move sometimes faster and sometimes slower than the sun (pp. 106–07[172]), but still sees this as the result of the 'contrary power', which is also invoked to explain the fact that Mars, Jupiter, and Saturn can be seen at all angles of divergence from the sun. Moreover the power is made to account for the phenomenon of planetary retrogradation as well (p. 110). We are required to believe that a power that causes Venus and Mercury to overtake and be overtaken by the sun is of the same nature as one that causes Mars, Jupiter and Saturn to be continually overtaken by it, and all the planets periodically to stop in their courses, go back on their tracks, stop again, and finally resume their courses. Cornford's fundamental error is to manufacture a connection (by means of the 'contrary power') between the movements of what we call the inner and the outer planets. There is not the slightest justification for this in the text, which specifically confines this power to Venus and Mercury. The movements of the other planets can perfectly well be represented (at least in the schematic fashion with which Plato is here concerned[173]) by assuming different speeds of revolution for them relative to the sun, but they are conceived as moving in separate orbits which really have nothing to do with the sun, except in so far as the latter by its uniform motion in a year provides a convenient unit for measuring the others. It is difficult to resist the suspicion that Cornford has allowed himself to be led astray by his own imagery of a moving staircase, which is anyway inconsistent with his idea of the physical model that he supposes Plato to have had in front of him, and which is not particularly apposite in an explanation of the separate circular orbits of Platonic astronomy.

There is no doubt that the basic picture Plato had in mind was of moon, sun, and planets revolving at different speeds round the central earth. In addition he knew that, because Venus and

Mercury are never seen far from the sun and complete their circuits of the zodiac in the same time as the sun,[174] the simple assumption of different speeds of revolution would not suffice to account for the characteristic behaviour of these planets. He therefore endowed them with 'the power opposite to that of the sun' which made them 'overtake and be overtaken' by it. This, of course, is not intended to be a scientifically accurate explanation of the phenomenon but is perfectly appropriate to the general tone of his story of creation. It is more difficult to account for the statement in 36d that 'he (the Demiurge) prescribed that the circles should move in opposite senses to one another' (κατὰ τἀναντία μὲν ἀλλήλοις προσέταξεν ἰέναι τοὺς κύκλους). If ἀλλήλοις refers to the seven circles of the Different (which is the most natural interpretation since these have just been mentioned in the last part of the previous sentence[175]), then it would seem that Plato has made a slip; for, whether we take it that it is Venus and Mercury or some or one of the other planets that is supposed to move in the opposite direction, the statement does not accord with the facts of observation. The error (if it is one) is not of a particularly heinous nature.[176] Given the facts that Plato was not an expert astronomer, that he is not writing an astronomical treatise, and that the whole concept of the celestial sphere and of the planets revolving in their orbits at different speeds was a novel one at that time and is, anyway, by no means easy to grasp, we should hardly expect a wholly correct explanation of a complicated phenomenon in the short section of the dialogue dealing with astronomical details. Knowing that later (38d) he was going to postulate a power contrary to the sun to explain the peculiar characteristics of the motions of Venus and Mercury, he might have supposed that he was foreshadowing this by the loose expression κατὰ τἀναντία (ἐναντίος being used in both passages), without realizing that, taken literally (and naturally), the words were not appropriate to the phenomenon he had in mind. On the other hand, it is possible (though perhaps unlikely) that ἀλλήλοις refers, not to the seven circles of the Different, but to the two main circles of the Same and the Different, which have also been mentioned in the previous sentence, but before the seven circles; in which case there is no

error and no problem.[177] The definite contrast between κατὰ
τἀναντία μέν . . . and τάχει δέ might be argued to support this, but
the contrast can equally well be interpreted as being between the
directions of movement of the seven circles and their speeds (cf.
Cornford, p. 80).

Timaeus goes on to explain (38e–39a) how, after being set in
their appropriate orbits, the celestial bodies were endowed with
living, intelligent souls and enjoined to move in circles of different
radii, those in the smaller circles moving more quickly and those
in the larger more slowly, but each one subject to the two main
motions (acting in opposite directions) of the Same and the
Different; the combined result of these two motions is that each
planetary body moves in a spiral (ἕλιξ). This spiral twist (relative
to the celestial sphere as a whole) may be envisaged from Fig. 10
(p. 122). A planet at P, after one twenty-four-hour revolution of
the circle of the Same, will not return exactly to P but to a point
P_1 some distance along BD, and after another revolution to a point
P_2 further along, and so on, until it reaches the point D after
describing a number of descending spirals of decreasing radius.
Similarly, from D (round the back of the diagram as it were) it
describes a series of ascending spirals of increasing radius (reaching
a maximum at the equator) and then again decreasing to B, all the
spirals being included between the limits of the tropics AB and DC.
Recognition of these spiral courses of the planets argues no small
astronomical insight on Plato's part, as there is no evidence that
any such notion was entertained before him. In fact, the Philolaic
system with its concentric orbits all in one plane (see above)
effectively precludes it. However the Eudoxan system, of course,
entails something similar.

Next, we are told that the god kindled a light (what we now
call the sun) in the second of the orbits near the earth, to provide
a means of measuring the comparative speeds of the eight
revolutions (i.e. of the Same and the seven circles of the Different),
and to familiarize living things with the concept of number. So
night and day came into being as the period of 'the one and most
intelligent revolution' (that of the Same), the month as the period
'when the moon, going round its own circle, catches up once

9

more (ἐπικαταλάβῃ) with the sun', and the year as the period when the sun completes its own circle (39b–c). Then follows the remark:

> The periods of the others [i.e. the five planets] men have not reflected upon, except a few of the many, nor do they give them names, nor investigate and make numerical measurements of them relative to each other; so that, as it were, they do not know that the wanderings of these constitute Time at all, since they are very complicated in number and of an amazing variety (39c–d).

The fact that men do not give these periods names means simply that there are no words like 'month' or 'year' for a complete circuit of the heavens of any of the five planets. Evidently in Greek astronomy as known to Plato very few planetary observations had been made, and this is consistent with the comparatively late recognition of the planets as such and the fact, for example, that we are told nothing of the periods assigned to them in the Philolaic system (see above); Eudoxus must have been one of the 'few men' to study them, but Plato's words support the idea that the data Eudoxus relied on to construct his system came from outside Greece (almost certainly from Babylonian astronomy, as we shall see).

We are then told that, despite the difficulties, it is possible to conceive of a 'perfect number' (τέλεος ἀριθμός, 39d) of days which constitute the 'perfect year' (τέλεον ἐνιαυτόν) or period when all the eight circuits (of the fixed stars, sun, moon, and planets) have completed their revolutions together. This recalls the 'perfect number' of Republic viii, 546b, which can hardly be described as anything other than number mysticism.[178] In the present context it has equally little astronomical significance,[179] although it is just possible that some reference is intended to the correlation of lunar and solar cycles, which was certainly being investigated in Plato's time; a period of years containing an integral number of lunar months was commonly known as a 'great year' (see pp. 75f.; 86ff.). This section ends with the statement that all the bodies that exhibit 'turnings' (τροπάς) in their paths through the heavens (δι' οὐρανοῦ) –

i.e. sun, moon, and planets, the courses of which are all confined, and so 'turn', between the limits of the tropics (see above) – were brought into being in order that the universe might be as similar as possible to the perfect, intelligent Being in respect of its ever-lasting nature (39e).

The Demiurge now sets about populating the universe with four kinds of living creature, corresponding to the four basic elements that Mind recognizes in the cosmos, the divine kind of the heavens corresponding to fire, the winged kind to air, the watery kind to water, and the dry-land creatures to earth (39e–40a). Previously Plato has had to anticipate the creation of sun, moon, and planets as divine beings because they were required to generate the concepts of number and Time; now he resumes the logical order of exposition. The divine kind (the stars) were fashioned 'mainly from fire', made circular (εὔκυκλον), set to follow the 'intelligence of the supreme' (i.e. to revolve with the circle of the Same), and 'distributed in a circle throughout the whole heaven' (νείμας περὶ πάντα κύκλῳ τὸν οὐρανόν, 40a). Each was endowed with two and only two motions, one rotatory on its own axis ('always thinking to itself the same thoughts about the same things'), and the other a forward motion 'governed by the revolution of the Same and the uniform' (40a–b). Such was the generation of the fixed stars (ὅσ'ἀπλανῆ τῶν ἄστρων), 'divine living creatures, which are eternal and always abide turning round uniformly in the same place'; 'but the bodies which exhibit turnings, and have an irregularity in this sense, came into being in the manner which was described previously' (40b).

There is no doubt that Plato makes a clear distinction here between the fixed stars on the one hand and the planetary bodies (including sun and moon) on the other; it is only the former, the generation and nature of which are described in detail here, that are specifically endowed with rotatory movement on their own axes, whereas the generation, nature and movements of the planetary bodies have already been described in the previous sections (38c–39b). Thus, there is really no justification for Proclus' assumption[180] that Plato meant to endow all the celestial bodies (including the planets) with an individual, rotatory movement, nor does the fact that each is a divine, intelligent being, possessed of a soul (and therefore capable of self-motion, since soul is the principle of motion – *cf.* Procl. 275*f*–276*a*) render this assumption necessary. The planets and the sun and moon have already been specifically stated to have two motions, that of the Same (which they share with the fixed stars) and that of the Different (which includes their own proper motions), the combined effects of which produce the spiral twist (see above). In addition, Venus and Mercury have a special power whereby they 'overtake and are overtaken' by the sun, and also all the planets exhibit the pheno-menon of retrogradation which there is no good reason to suppose Plato did not recognize (*cf.* Cornford, p. 110), although in his designedly schematic account there has so far been no specific mention of it (but see below). Hence there is plenty of scope for the soul of each planetary body to exercise its own prerogative of self-motion, without attributing to it also the property of axial rotation. On the other hand, the fixed stars are affected only by the overriding movement of the Same, so that it is logical to attribute to them, as the possessors also of divine souls, an addi-tional motion that can only be one of rotation on their own axes since their only translational motion in the sky is that of the daily revolution of the celestial sphere.

There follows the statement (40*b*–*c*): 'Earth (the Demiurge) designed to be our nurse, and also, as it is packed round the axis stretched through the whole (heaven), guardian and creator of both day and night, the first and oldest of all the gods that have come into being within the universe.'[181] This sentence has been the cause of considerable discussion by commentators, because (*a*) there has been doubt about the form and meaning of the word here translated 'packed round', and (*b*) an apparent quotation of it by Aristotle has been interpreted as implying that he attributed to Plato the notion of a moving earth. To take these points in order: (*a*) Burdach's article (see note 181) has cleared up a number of misconceptions. There is evidence that even by the fourth century BC the distinction between two originally separate Greek stems (but probably, according to Burdach, both derived from a single pre-Greek 'vel-root') with originally separate meanings was becoming blurred, and the numerous variant spellings (εἰλ(λ)ομένην, εἰλ(λ)ομένην, εἰλουμένην, εἰλουμένην, ἰλλομένην, ἰλλομένην, κ.τ.λ.) found in the mss of Plato, Aristotle, and their ancient commentators testify to ambiguities in both the form and meaning of the word. Therefore, philological and textual considerations alone are of limited value in deciding what Plato actually intended to convey by his use of it. (b) The passage in Aristotle occurs in *De Caelo* ii, 13, 293b30 where, in a discussion as to whether the earth is to be thought of as moving or stationary, it is stated: 'Some say that, although it lies at the centre, it coils and moves round the axis stretched through the whole (universe) as has been written in the *Timaeus*.' A little later on (296a26) there is a reference to 'those who place (the earth) at the middle but say that it coils and moves round the central axis'.[182] If the second passage is a reminiscence of the first, and if the reference to the *Timaeus* in the first is taken as Aristotle's own words, then it would seem that he here attributes to Plato the idea that the earth is not immobile.

The only motion that can be envisaged for it is rotation round its own axis, since the whole gist of Aristotle's arguments in this chapter shows that he divides those who have given opinions about the earth's position and motion or lack of it into two broad categories, those who hold that it is not at the centre of the universe but circles round the centre like a planet (the Pythagorean view – see above), and those who hold that it is at the centre (cf. *De Caelo* 293a18ff. and 293b18ff.). Yet if the earth is endowed with axial rotation, day and night will not be produced by the revolution of the Same, as was specifically stated at *Tim.* 39b–c (see above, p. 129), and the whole astronomical system presented in the dialogue, with the revolution of the Same as the overriding factor that affects also the movements of sun, moon, and planets, will be upset. Moreover, a rotating earth is totally inconsistent with the astronomical notions in the other dialogues, and even with other passages in Aristotle where it is clear that he is referring to Plato's astronomy.[183]

The ancient commentators were fully aware of the difficulty; Simplicius after discussing various views (*in De Caelo* 517ff.) suggests that καὶ κινεῖσθαι in the first passage may be a later interpolation. This would certainly remove the difficulty but, as Heath points out (*Arist.*, p. 177), this explanation is ruled out by the fact that Aristotle prefaces the second passage (where exactly the same expression is used) by the words 'as we said before' (καθάπερ γὰρ εἴπομεν, *De Caelo* 296a25). Other suggestions that have been made include taking the words 'as has been written in the *Timaeus*' as referring only to the last part of the sentence, i.e. the phrase 'round the axis stretched through the whole' (this suggestion goes back to Thomas Aquinas – *cf.* Dreyer, p. 78) or supposing that Aristotle is here deliberately misrepresenting Plato's views.[184] None of this is convincing. Cornford, who takes it as certain that Aristotle definitely meant to impute to Plato the idea of a rotating earth at the centre of the universe, argues with his customary ingenuity that a rotating earth is in fact implicit in the astronomy of the *Timaeus*, on the ground that, since the World-Soul is described as being 'everywhere inwoven from the centre to the extremity of heaven and enveloping the heaven all round on the outside, revolving upon itself' (*Tim.* 36e), this must affect the earth as well. 'If this movement alone existed,' says Cornford (p. 130), 'it would be indistinguishable from rest. There would be no change in the relative positions of any parts of the world's body, and there would be no day and night.' Hence, he argues, Plato must have assumed a rotation of the earth in the opposite direction to and exactly cancelling out the revolution of the Same which produces the diurnal movement of the heavens; so that in effect, relative to the fixed stars, the earth *is* stationary at the centre of the universe, but an observer in outer space would see both the stars and the earth revolving in opposite directions.[185] Cornford buttresses his argument by pointing out that since the earth is described as 'the first and oldest' of the gods in the universe, it must be endowed with a soul (like all the rest of the divine, heavenly bodies – see above) and therefore it would be appropriate for it to exhibit some form of motion.

Despite the ingenuity of Cornford's arguments, there can be little doubt that the weight of the evidence is against imputing to Plato the concept of a rotating earth. Not only does this concept accord ill with the astronomical notions apparent in the *Republic*,

Phaedo, and *Phaedrus* (see above), but in the *Timaeus* itself the 'perfect year' is defined as the period when all *eight* circuits (i.e. of the fixed stars, sun, moon and five planets) have completed their revolutions together (see above, p. 130), not nine as would have to be the case if the earth itself was in motion (*cf.* Proclus, *in Tim.* 281*f*). Moreover, the *Epinomis* specifically mentions the eight 'powers' or circuits of the heavens (986*a–b*, ὀκτὼ δυνάμεις τῶν περὶ ὅλον οὐρανόν: *cf.* 986*e*; 987*b*; 990*a*, τῶν ὀκτὼ περιόδων τὰς ἑπτὰ περιόδους),[186] and, in the discussion about the relation of the souls of the heavenly bodies to the actual moving bodies themselves (in *Laws* x, 898*c*–899*b*), it is only sun, moon, and stars (including the planets) that come into question, not the earth itself, as might be expected if, as well as being endowed with a soul, it was also a moving body. In fact, Cornford's whole case[187] for a rotating earth in Plato's astronomy is designed to account for Aristotle's remark in the first of the passages cited above (*De Caelo* 293*b*30, ἴλλεσθαι καὶ κινεῖσθαι) which he takes to mean that 'beyond question' Aristotle interpreted Plato here as believing in an earth with a ' "winding" motion round the axis of the universe' (p. 121).

This, however, as Cherniss emphasizes, (*op. cit.*, p. 551), is not the only way to read the sentence; it would be equally consistent with the Greek to take the words 'as has been written in the *Timaeus*' to be still part of the quotation introduced by 'Some say . . .' – in which case it is not Aristotle himself who imputes a moving earth to Plato, but the anonymous 'some'. (These need not be Pythagoreans whose opinions on the motion of the earth have been discussed in the previous section.) In order to support their own views they chose to interpret the comparatively rare and ambiguous word εἰλλ-(ἰλλ-)ομένην (whichever was the original spelling) in *Tim.* 40*b* as implying motion. This way of taking the passage effectively eliminates the need for attributing to Plato the notion of a rotating earth, for which there is no good evidence (*pace* Cornford), and would seem preferable to one that entails the assumption of Aristotle's acquiescence in such a notion without any of the further discussion that such an important innovation in Plato's thought might have been expected to require. The one criticism that might perhaps be directed against it is that it does not seem to have occurred to any of the ancient commentators.[188]

If we discard the idea that Plato envisaged a rotating earth, and accept Burdach's view that εἰλλομένην (ἰλλομένην) means 'packed round' without any connotation of actual movement (as Proclus understood it, *in Tim*. 281*d*), then it becomes possible to give a satisfactory explanation for Plato's choice of words in this passage.[189] Earth is described as being 'guardian and creator' of day and night because, being divine and endowed with a soul, it can resist the all-pervasive influence of the World-Soul which is the cause of the diurnal rotation of the whole universe (i.e. the revolution of the Same). Cornford is undoubtedly right in insisting that this must affect the earth as well. Were the earth an inanimate mass, it would share in the uniform revolution of the whole heaven from east to west and there would be no day or night. However since it is a divine, living being it possesses powers of resistance which it exercises continuously (like an ever-vigilant guardian or watchman, φύλαξ), and by exerting itself to remain stationary 'creates' night and day, night directly by means of its own shadow and day by virtue of the contrast with night.[190] Where Cornford goes astray is in supposing that Plato envisaged an actual motion of the earth. As Cherniss says (*op. cit.*, p. 554), if this were the case Plato would hardly have used the term εἰλλ-(ἰλλ-)ομένην in preference to such unambiguous words as περιφερομένην, κυκλουμένην or στρεφομένην; but since soul was normally the principle of motion, he might well have avoided using these for the resistant power exercised by the earth's soul, which in this instance kept the earth stationary (pp. 555–56).

So it is not in fact Aristotle who imputes the notion of a rotating earth to Plato, but a certain school of thought which interpreted (incorrectly) the *Timaeus* passage in this manner in order to support its own view of a centrally placed, rotating earth. Aristotle tells us nothing further about this school (referred to simply as ἔνιοι, 'some'), but it has been plausibly conjectured (e.g. Cherniss, pp. 551*ff*.) that the reference is to Heraclides Ponticus, a member of Plato's Academy, who certainly put forward the idea that the apparent diurnal motion of the heavens is caused by the earth's rotation in the opposite direction (from west to east) at the centre of a stationary sphere of the fixed stars, and who is mentioned both by Proclus[191] and by Simplicius[192] in connection with the reference to the *Timaeus* in *De Caelo* 293*b*30.

The astronomical section of the Timaeus ends (40*c–d*) with an enumeration of celestial phenomena which, says Timaeus, it is useless to discuss without the consideration of actual models.[193] The phenomena include 'circlings and juxtapositions' of the celestial bodies,[194] 'retrogradations[195] and forward movements', conjunctions and oppositions, and finally the times when they pass in front of each other and are hidden from our sight and then reappear, creating alarm and portents of the future 'for men who cannot calculate' (τοῖς οὐ δυναμένοις λογίζεσθαι) – the last phenomena, of course, refer to eclipses. From this list (although not expressed in technical terms) it is clear that most of the problems that occupied the attention of later astronomers were already recognized by the time that the *Timaeus* was written.

In the *Laws* Plato, in the guise of the Athenian stranger, takes up again the theme of the place of astronomy in an ideal education. As in the *Republic*, he stresses the great importance of mathematics as a discipline, not merely because of its general applicability to many other branches of knowledge (this aspect receives greater approval in the *Laws* than in the *Republic*; contrast *Laws* 819*c* with Socrates' rebuke of Glaucon for stressing the utility of the sciences, *Rep.* 526*d* and 527*d*), but also because the study of it quickens the intellect and sharpens the reasoning powers (*Laws* 747*a–b*; *cf. Rep.* 525*c–d*; 526*b*). In Book vii we are told that the secondary education (i.e. up to sixteen years old, *cf.* 809*e*–810*a*) of the children of free citizens ought to include enough astronomy to enable them to understand calendaric problems for the proper regulation of the city's life, e.g. for the fixing of seasonal sacrifices and festivals (809*c–d*). Only the few who are going to hold high office need know all, but every free man ought to know a little about three subjects, arithmetic, mensuration, and 'how the stars have come by nature to carry out their courses relative to each other' (817*e*). The importance of mathematics and astronomy in fostering the divine element in man's make-up is emphasized in 818c, where it is stated that the man who cannot count and who is 'unversed in the revolutions of sun and moon and the other stars' can never attain his highest development.

The Athenian goes on (820e–821b) to deplore the scandalous state of astronomical studies at the present time when no proper research is being carried out and practically all Greeks hold erroneous notions about the sun and moon, namely, 'that they never traverse the same path, nor do certain other stars as well as these, which are called planets' (literally 'wanderers'). Cleinias agrees that he himself has often observed that 'the Morning Star and the Evening Star and certain others never traverse the same path but wander in every direction, and that we all know that the sun and moon do this at some places'[196] (821c). The Athenian says that this is precisely why he advocates that everyone ought to have at least a little knowledge of astronomy, to know enough not to make such blasphemous statements about the heavenly bodies. On being pressed to explain, he states that he has not long heard about these things, but that they are not too difficult to explain (821e); the explanation (822a–b) after this elaborate build-up comes as something of an anticlimax. It is all wrong to believe that the moon and the sun and the other planets 'wander', for the truth is the exact opposite: 'each of them pursues the same path, not many paths but always one circular path, although it appears to be borne in many paths'. It is also wrong to suppose that the swiftest of them is the slowest and vice versa.

This statement about the paths of the planets has been held (a) to be inconsistent with the planetary movements depicted in the Timaeus (e.g. 36b–d), and (b) to constitute evidence that Plato did in fact change his mind about the role of the earth, as Plutarch said he did (see p. 235 note 145), and made it rotate on its own axis (this is held to explain the statement that the planets have only 'one circular path'). It is even held to prove that he came to believe in an earth rotating round the sun.[197] A. Boeckh in his Untersuchungen über das kosmische System des Platon, Berlin, 1852, over one hundred years ago convincingly refuted the last two suppositions, and Cornford (p. 90) has shown that (a) is untenable. He rightly points out that 'to be carried in many paths' (πολλὰς ὁδοὺς φέρεσθαι, 822a) means to traverse different tracks at different times (this is what the 'wanderers' *seem* to do) and cannot be interpreted as a composite motion resulting from other separate motions. The fact that Plato wants to emphasize here is that each planet has a fixed path from which it does not arbitrarily deviate. This one path is loosely called 'circular' because the planets do circle the earth in a geocentric universe, just as, after mentioning the spiral twist at *Tim.* 39a, he goes on to speak of the respective 'circles' of moon and sun (39c). As Cornford remarks (p. 91), 'If a man ascends a spiral staircase, he is not straying from one path to another. His position at any moment can be calculated. . . .' Thus Plato is not denying composite motion to the planets,[198] but stressing, in accordance with his general insistence on the mathematical side of astronomy, that their courses are regular (and hence, in theory, predictable) and that they are not the erratic 'wanderers' of uninformed opinion. In fact, far from being directed against Eudoxus,[199] Plato's remark is more plausibly interpreted as a tacit approval of the type of approach to the problem of planetary movements illustrated by the Eudoxan system,[200] which could hardly have been elaborated without the use of diagrams.

The second statement, relating to the misconception that the swiftest body is in reality the slowest and vice versa, is evidently directed against Democritus. Lucretius tells us (see pp. 83f.) that Democritus regarded the sphere of the fixed stars as revolving with the greatest velocity, and the other planets and the sun and moon as moving more slowly in proportion as their distances from the outermost sphere increase, the moon being the slowest of all; but to our eyes, the moon seems to move the quickest since it returns to the same zodiacal sign in the shortest period (for the confusion regarding the different planes of revolution, see above, *loc. cit.*). Plato agrees that the revolution of the fixed stars is the fastest (the outermost whorl of the spindle of Necessity – see above, p. 111), but (more correctly) treats separately the seven revolutions of the planetary bodies (i.e. the seven inner whorls of the spindle (p. 111) or the seven circuits comprised in the circle of the Different (p. 128) the speeds of which are made proportional to their radii (or their distance from the earth). The moon, which moves in the smallest circle and nearest the earth, is the fastest in an absolute and not merely a relative sense.

In Book x Plato animadverts against the view that the four elements, fire, water, earth, and air, came into being simply 'by nature and chance, and none of them by design' (φύσει πάντα εἶναι καὶ τύχῃ φασίν, τέχνῃ δὲ οὐδὲν τούτων, 889b), that the earth, sun, moon, and stars are entirely 'soulless' (ἄψυχοι), and that the whole universe and all it contains is not the result of the workings of Mind (νοῦς), or a god, or any design, but was generated by nature and chance or (as he also puts it) 'necessarily by chance' (κατὰ τύχην ἐξ ἀνάγκης, 889c).[201] The obvious reference here is to the mechanistic universe of the Atomists, in which the atoms come together through the action of the primeval vortex 'by necessity' (δι'ἀνάγκην – see above, p. 79). However, mention of the four elements, which played no special part in atomist theory, shows that Plato was also thinking of other Ionian physical theories of an equally non-teleological kind – e.g. according to Parmenides Necessity guides the heavens (see above, pp. 50f.), and in Empedoclean physics both Necessity and Chance play a large role (cf. Guthrie, op. cit., pp. 161–62). The idea that such inanimate 'causes' could act as the mainspring for the entire universe was anathema to Plato (cf. Socrates' remarks in the Phaedo – above, p. 95). 'Soul' (ψυχή), he insists, is the governing principle of the universe and the cause of all motion, and it existed before body which is ruled by it (892a; 896b–d). There is good soul and bad soul in the universe (896e; 897) and it is the former that controls the movements of sun, moon, and stars. Since everyone can see that the sun is a body, soul can only be recognized by the mind – it is a νοητόν (898c–e). Moreover, the good soul or souls of the heavenly bodies, which produce the years and months and seasons, must be regarded as gods 'whether they direct the whole universe by inhabiting these bodies, as living entities, or in whatever other manner they act' (899b). Thus, for Plato, a belief in gods is a *sine qua non* for the correct understanding of the workings of the universe.

A discussion of the best means of inculcating this belief upon the minds of the members of the most important body in the state, the 'nocturnal council' (νυκτερινὸς σύλλογος, 409a; cf. 951d; 961a–b), forms the subject matter of the last part of the last book of the Laws (xii). Here it is stated that there are two main ingredients in such an education: (a) the knowledge that soul is the oldest and most divine thing in the universe, and (b) knowledge of the orderly movements (controlled by Mind) of all the celestial bodies, which cannot be contemplated without realization of their divinity (966d–e). The idea that those who devote themselves to astronomy and related subjects (ἀστρονομίᾳ τε καὶ ταῖς μετὰ ταύτης ἀναγκαίαις ἄλλαις τέχναις) become atheists is entirely wrong (967a). Astronomers themselves must take responsibility for this erroneous belief, because, although certain of them have said that Mind controls everything in the heavens, they have failed to realize that soul is older than body, and have stated that 'the moving objects in the heavens are full of stones and earth and many other soulless bodies' (τὰ κατ' οὐρανὸν φερόμενα μεστὰ εἶναι λίθων καὶ γῆς καὶ πολλῶν ἄλλων ἀψύχων σωμάτων). They then assign these as causes of the whole universe (967b–c – an obvious reference to Anaxagorean doctrines; cf. above, pp. 94–5), hence the bad reputation of philosophical enquiry among the poets (e.g. the satirical picture of Socrates' 'think-shop' in Aristophanes' Clouds). In fact, no man can attain to a sound religious understanding (γενέσθαι βεβαίως θεοσεβῆ) without being convinced of the truth of the above two dicta, namely, the priority of soul, and the presence of Mind in the stars; only such a man, who has also mastered the necessary preliminary studies, i.e. especially the whole field of mathematics, is fit to hold high rank in the ideal state (967d–968a).

Finally, in this examination of Plato's astronomical ideas, we come to the Epinomis, a dialogue that employs the same characters as the Laws, is sometimes cited as the thirteenth book of that work,[202] and is evidently intended as an appendix to it, since it deals with the question 'Where is Wisdom to be found?' (cf. Epin. 973b, τί ποτε μαθὼν θνητὸς ἄνθρωπος σοφὸς ἂν εἴη), a question that might be thought to arise naturally from the more generalized discussion at the end of the *Laws* as to the best type of education for members of the nocturnal council. The genuineness of the dialogue has been much disputed by modern scholars, despite the fact that ancient testimony is almost unanimous[203] in accepting it as Plato's work. No attempt can be made here to detail the arguments for and against its authenticity.[204] Suffice it to say that the doctrinal objections urged by Cherniss[205] are not nearly so cogent as Tarán would have us believe,[206] and it is difficult to resist the suspicion that dislike of its subject matter (e.g. its recommendation of an astral religion, not excluding a demonology that would lend itself readily to the tenets of astrology – see below) has been the main reason prompting scholars to disown it as Plato's. This was certainly the case with Proclus who, as Taylor points out,[207] can cite no testimony for his opinion. The Master must be saved from his own indiscretions! However, even those scholars who, knowing better than Plato what he should have written, deny the authenticity of the *Epinomis*, agree that its author must have been closely connected with the Academy, and since the astronomy proper in the dialogue is fully consistent with that found in the rest of the Platonic corpus, it merits consideration here.

The keynote of the dialogue, the insistence on the divinity of the heavens as a fit subject for human worship because of the gifts it bestows on men,[208] is struck early. The 'Universe' (οὐρανός, also called κόσμος or ὄλυμπος, 977*b* – *cf.* above, p. 118) is the greatest god and responsible for all benefits to mankind including the best of all, the gift of number (ἀριθμός), for by making the stars carry out their evolutions 'in itself' (ἐν αὐτῷ) it creates the seasons and provides nourishment for all (977*a–b*). Without the knowledge of number man is nothing, possessing neither wisdom nor art (977*c–e*). The Universe never ceases to teach men how to count by the alternation of day and night, by the fifteen-day periods of the waxing and waning moon, and by the grouping of the months into a year (978*d*–979*a*).

There are five kinds of body, viz. fire, water, air, earth, and aether, and two different classes of visible being, one the earthy type (γήινον) consisting mainly of earth and solid matter but containing also portions of all the other elements (this class includes men, animals, and all vegetable matter), and the other consisting mainly of fire but also with an admixture of the other elements. All celestial objects ('the divine race of the stars', θεῖον γένος ἄστρων) belong to the second class, which is endowed with the most beautiful body and the best and most fortunate soul (981d–e). The order and regularity of celestial movements (as contrasted with the disorder, ἀταξία, of life on earth) is sufficient proof of the intelligence (τοῦ φρόνιμον εἶναι) of this second class of being (982b), and, in the same way, the proof that the stars and the whole celestial procession (διαπορείαν) possess Mind (νοῦν ἔχειν) is to be seen in the uniformity and continuity of their actions, which are carried out 'according to plans made an enormously long time ago', so that they do not 'change their minds this way and that way, doing different things at different times, or wander about and change their orbits' (οὐ μεταβουλευόμενον ἄνω καὶ κάτω, τότε μὲν ἕτερα, ἄλλοτε δὲ ἄλλα πρᾶττον, πλανᾶσθαί τε καὶ μετακυκλεῖσθαι – 982c–d). Yet men in their foolishness have taken this very regularity and uniformity as proof that the stars have no soul – an impious belief and one that can be countered by a proper appreciation of the beauty and majesty of the circumvolutions (χορείας – the same word and the same sense as *Tim.* 40c, see above) of the stars (982d–e).[209]

Further proof that the stars are endowed with souls (ἔμψυχα) is afforded by their size; they are not as small as they appear, but are in fact 'each of gigantic mass – and this must be believed for it is established by sufficient proofs' (ἀμήχανον ἕκαστον αὐτῶν τὸν ὄγκον – πιστεῦσαι δ'ἄξιον· ἀποδείξεσιν γὰρ ἱκαναῖς λαμβάνεται, 983a).[210] The sun is larger than the whole earth and each star is of a huge size; only a god could cause such huge masses to move always in the same periods as they now move and endow them with soul, and the whole universe and all the stars must be possessed of soul in order to carry out their courses in years, months, and days (983a–c). If anyone thinks that 'whirlings' (ῥύμας – an obvious reference to the primeval vortex theory: cf. above, pp. 52–3; 57) or natural substances or any such thing are the causes of bodies, he is wrong; there are only two kinds of existence, soul and body (but there are many kinds in each group, all differing from one another and from their counterparts in the other group), and of these soul has intelligence (ἔμφρον) and governs body which has no intelligence (ἄφρον). It is rank stupidity to suppose that the objects in the heavens (τὰ κατ'οὐρανόν) came into being from anything other than soul and body (983c–e),[211] and these same objects must be either gods or the images of gods created by the gods themselves, and as such worthy of the utmost reverence (984a).

Plato then goes on to postulate three other types of divinity intermediate between the visible celestial gods (whose chief constituent is fire – see above) and terrestrial life (chief constituent earth); these three are constituted from aether,[212] air, and water respectively by the action of soul, which fills the whole universe with different types of living being (πάντα δὲ δημι- ουργήσασαν ταῦτα ψυχὴν ζῴων εἰκὸς ὅλον οὐρανὸν ἐμπλῆσαι, 984c; cf. 985a ad fin.). The conventional gods of mythology (Zeus, Hera, and the rest) may be placed wherever one likes, but the greatest and most honourable of the visible gods are undoubtedly the stars; after these and inferior to them come the daemons (δαίμονες), then the gods of air (who deserve special honours because of their role as interpreters and intermediaries), and finally the demi-gods of water – these last-named may sometimes be dimly seen by mortals, but the daemons and the airy kind are invisible (984d– 985b).[213] The lawgiver 'who has even the smallest amount of sense' (ὅστις νοῦν κέκτηται καὶ τὸν βραχύτατον, 985c) will not try to convert his state to new worships at the expense of the orthodox ones; but it is thoroughly undesirable that the visible gods should not receive their due share of festivals and sacrifices (985e).

There is no doubt at all that the existence of these star gods and of the five classes of being (three divine, one semi-divine, and one terrestrial) is regarded as incontrovertible fact (cf. 985c, τούτων δὴ τῶν πέντε ὄντως ὄντων ζῴων, and 985d, τοὺς δὲ ὄντως ἡμῖν φανεροὺς ὄντας θεούς), and one can sympathize with those scholars who prefer to regard such doctrines as un-Platonic; certainly this description of a god-haunted world with its denizens of aether and air acting as go-betweens between mankind and the uppermost

gods (985b) seems a far cry from the austere rationality of the
earlier Plato, and might seem better suited to the astrologically
orientated world of later antiquity.[214] Bouché-Leclercq has shown
how even the doctrines of the *Timaeus* were seized on and re-
interpreted by the Neo-Platonists and later astrological writers to
buttress their arguments for a world dominated by astral influences and to support the Stoic view of the microcosm, man, as
a diminutive replica of the macrocosm, the universe;[215] the ideas
expressed in the *Epinomis* (which, curiously, Bouché-Leclercq
does not mention) might be thought to be even more apposite in
this context. Yet, although these ideas are of just the type to suit
the background of astrological thinking and provide a plausible
philosophical basis for it,[216] there is no explicit statement in Plato's
works of any definite astrological belief – e.g. there is nothing
about the predictability of terrestrial events from the positions of
the celestial bodies and nothing about the changing influences of
the latter on mankind according to their different configurations,
which are two of the basic premises for astrology. Our judgment
as to how sympathetic Plato was to such ideas must be closely
connected with our opinions on how much he was influenced by
oriental modes of thought; for astrology originally came into
Greece from the East, from Babylonia.[217] A case can be made for
supposing that he may have been more influenced by eastern ideas
than is generally supposed;[218] certainly there is no reason why he
should not have had ready access to them, probably through
Eudoxus who undoubtedly used Babylonian astronomical results
and evidently knew about Babylonian astrology, since, according
to Cicero (*De Div.* ii, 87), he rejected that branch of it known as
'genethlialogy', i.e. horoscopic astrology.[219]

   To return to the astronomy of the *Epinomis*: there are eight
'powers' (δυνάμεις) in the heavens, one of the sun, one of the moon,
one of the (fixed) stars,[220] and five others; all these are gods and
should be accorded due honours (986b–c). These 'powers' designate the circular orbits of the celestial bodies, as is made clear by
986e where it is stated that the fourth 'revolution and orbit'
(φορὰ καὶ διέξοδος) and the fifth keep pace with the sun (that is the
fourth and fifth of the five planets, counting from the outermost,

Saturn). These are Mercury and Venus which already in the *Timaeus* (38*d*) have been described as moving with the same speed as the sun (*cf.* above, pp. 123; 127–28); of the three 'the one that has sufficient mind leads the way' (τὸν νοῦν ἱκανὸν ἔχοντα ἡγεῖσθαι), i.e. the sun. Plato then explains that, while one of the sun's companions (Venus) has a name in Greek, ἑωσφόρος, 'Morning Star' (and also – as we see later – ἕσπερος 'Evening Star'), the other has no name, because the first observations were made in foreign (βάρβαρος) countries such as Egypt and Syria where, because of their favourable climates, observation of the stars was much easier and had been carried on for a very long time, and from which such knowledge had come to Greece (986*e*–987*a*). These foreigners gave to the planets the names of deities – hence 'the star of Aphrodite' (Venus) and 'the star of Hermes' (Mercury).[221] Here we have the earliest explicit statement in Greek sources of the debt of Greek astronomy to eastern civilizations. It was to be repeated by Aristotle (*De Caelo* 292*a*8 – see below). As we shall see, this is undoubtedly true in the case of the Babylonians, but we now know that Egyptian astronomy remained at a very elementary level until Hellenistic times, when it was strongly influenced by Greek methods and concepts, which were applied largely for astrological purposes, and that the early Greek astronomers had little to learn from this quarter.[222] The profound impression made by the obvious antiquity of the Egyptian civilization and its colossal monuments seems to have been the main reason for the attribution to the Egyptians of proficiency in all branches of knowledge from time immemorial. It was constantly made by Greek writers from Herodotus (e.g. Book ii) onwards; Plato himself acquiesced in this, as *Tim.* 22*ff.* shows.

The *Epinomis* goes on to speak of three other orbits moving with those of the sun and moon 'towards the right' (ἐπὶ δεξιά, 987*b*),[223] and of the eighth orbit which moves in the opposite direction to all these and might properly be called the 'cosmos' (i.e. the sphere of the fixed stars), and which 'guides the others, at least as it might seem to mere mortals who know little of these matters' (ἄγων τοὺς ἄλλους, ὥς γε ἀνθρώποις φαίνοιτ' ἂν ὀλίγα τούτων εἰδόσιν, 987*b*). These words (as well as the statement in

Laws 822*a* that each planet has only one circular path – see p. 138) have been taken to imply that Plato is denying that the diurnal rotation of the sphere of the fixed stars affects also the revolution of the other celestial bodies (this is the overriding influence of the circle of the Same in the *Timaeus*), and that consequently he must be thinking of a moving earth with a stationary sphere of the fixed stars.[224] However, as has been shown above, a moving earth is incompatible with Platonic astronomy, and the passage in the *Laws* can be explained satisfactorily without such an assumption. Similarly, in this passage a more natural interpretation is to take ἀνθρώποις (without the article which would be almost necessary for Des Places' translation) as meaning 'mankind in general', give γε its full force, and obtain the sense 'at any rate in the eyes of mankind who (naturally) can know little of these matters'. This would be fully consonant with the warning as to the limitations of human knowledge with which Timaeus prefaces his account of the universe (*Tim.* 29*c–d*), and fits in well with the words that immediately follow in the *Epinomis* (987*c*, ὅσα δὲ ἱκανῶς ἴσμεν, ἀνάγκη λέγειν καὶ λέγομεν, i.e. but what we know reasonably well, we are bound to state).[225] The three remaining planets are then mentioned, Saturn being the slowest, then Jupiter, and after it Mars which has the reddest colour of all. Thus the order of the celestial orbits is (counting from the earth outwards) moon, sun, Venus, Mercury, Mars, Jupiter, Saturn, and fixed stars; this is the same order as in the *Republic* and the *Timaeus* (*cf.* p. 126).

The geographical position of Greece is now extolled as being the best for the development of virtue, and even though the Greeks, because their summer climate is less favourable than that of other regions (another reference to the borrowing of astronomical data from abroad – see above), have acquired knowledge late, yet whatever they take from foreigners they improve on (987*d*). The importance of number (988*b*), of the concept of soul (988*d*), and of the virtue of piety (εὐσέβεια, 989*b*) is once again emphasized, and this leads to the emphatic statement that, surprising as it may seem (σχεδὸν μὲν οὖν ἐστιν ἄτοπον ἀκούσαντι), the one subject that is best suited to provide the ideal education for the highest ranks in the state is astronomy (989d–990a). Your true astronomer (τὸν ἀληθῶς ἀστρονόμον) must needs be the wisest of men – 'not the man who practises astronomy after the fashion of Hesiod and all such, observing settings and risings, but one who observes the seven orbits out of the eight', a difficult subject not given to every nature to grasp (990a–b). The sentiment here is of a piece with Plato's recommendation to astronomers to concentrate more on the theoretical side of their subject and work out a mathematical scheme which would account for the movements of the sun, moon, and planets (see above, p. 106); this, and not the mere accumulation of imperfectly understood observational data, is what gives astronomy its value as an educational discipline.

The moon completes its period the quickest, bringing the month and the full moon; then the sun must be studied, which brings the solstices (τροπὰς ἄγοντα), and the bodies (i.e. Venus and Mercury) that orbit with it (990b).[226] The other planetary orbits are too difficult to discuss without a thorough preliminary training from boyhood upwards in various subjects, of which the first and most important is the study of pure number (ἀριθμῶν αὐτῶν ἀλλ'οὐ σώματα ἐχόντων, 990c); then comes geometry and then stereometry or solid geometry (990d; this is the same order as in *Republic* vii – see above, pp. 102–03). The accuracy of the celestial movements is emphasized (ὡς ἀκριβῶς ἀποτελεῖ πάντα τὰ κατ'οὐρανὸν γιγνόμενα, 990c) and it is pointed out that understanding of these, together with the knowledge that soul is both older and more divine than body, should lead to a proper appreciation of the saying that 'everything is full of gods' (θεῶν εἶναι πάντα πλέα, 990d).[227] Finally, we are told that 'every diagram and system of number, every composition of harmony and the revolutions of the stars must necessarily be revealed as a single unity of all' to the man who has had the right training, for there is one natural bond (δεσμός) between all these things (991e–992a). The dialogue ends with an exhortation to the nocturnal council to acquire such wisdom by means of the studies enumerated.

We see, then, that, as regards astronomy, the *Epinomis* contains nothing that is inconsistent with the earlier dialogues but develops

ideas that are already implicit in them, particularly concerning the
divinity of the celestial bodies and the importance of astronomy
as a mental discipline and also as a means of attuning the human
mind to the highest thoughts it is capable of, viz. the contemplation of the structure of the divine universe. The impression one
receives is of a gradually increasing awareness on Plato's part as
he grew older of the problems of astronomy and of the philosophical import of the solutions proposed. He stands on the
dividing line between the old pre-scientific notions of the Pre-
Socratics and the new scientific (and methodologically far-
reaching) concept of the celestial sphere as a mathematical
extension of the spherical earth. As we have seen, it was the last
decades of the fifth century and the beginning of the fourth that
saw the development of this concept, and it is just this very period
that forms the historical setting to most of the dialogues. Without
subscribing to Taylor's thesis that the astronomical views expressed
in the *Timaeus* are only those of the Pythagorean school and not
of Plato himself (see p. 231 note 118), nevertheless one can hardly
doubt that he was strongly influenced by Pythagorean notions,
and not least in the field of astronomy. He was also aware of the
'new astronomy' based on the new concept of the celestial sphere,
and it was the difficulty he found in trying to reconcile Pythagorean and the 'new astronomy', coupled with the fact that mathematical astronomy was not a field in which he himself was
particularly skilled, that is responsible for the tentativeness, not to
say vagueness, of his astronomical descriptions, which attempt an
unsuccessful compromise between the two systems.

   Comparison of the passages in *Republic* x and the *Timaeus*
(discussed above) reveals clearly that his astronomical knowledge
had undergone an unmistakable development in the thirty odd
years between the writing of the two. The circles of the Same and
the Different (inclined at an angle to each other) in the *Timaeus*
evidently represent respectively the sphere of the fixed stars
(responsible for the diurnal rotation of the whole heaven) and the
band of the zodiac in which sun, moon, and planets have their
individual motions; this is a considerable improvement on the
naivety of the picture of the spindle of Necessity in *Republic* x

with its eight whorls, which gives the impression of being strongly influenced by Pythagorean astronomical ideas (pp. 113*f.*). After the *Timaeus*, Plato makes no further attempts to give a detailed account of his astronomical beliefs, but instead we see how he began to place more and more emphasis on astronomy as an educational discipline. By 350 BC Eudoxus had probably put forward his theory of concentric spheres to try to account for the apparently irregular movements of the planets, which, of course, was based on the concept of the celestial sphere (see below), and it is difficult to believe that Plato was totally unaware of Eudoxus' work. It seems likely that while Plato did not have the mathematical competence to understand the theory completely, he nonetheless realized that it went closer than the Philolaic scheme to accounting for the observed phenomena. Moreover, the fact that Eudoxus' spheres were abstract conceptions, and that he was in a sense going behind the actual visible phenomena of planetary variations to postulate a series of ideal, circular motions, obviously fitted in excellently with the spirit behind Socrates' demand (*Rep.* vii, 530*b–c*) for the prosecution of astronomical research by means of problems rather than mere observation.

CHAPTER VI

EUDOXUS

EUDOXUS OF CNIDUS (in Caria on the south-west shore of Asia Minor, some way south of Miletus) is one of the key figures in the history of ancient astronomy; mathematician, astronomer, geographer, and philosopher, who wrote also on music and medicine, he was a polymath of considerable range, but unfortunately nothing of his work survives entire and the little we have is derived from quotations in later writers.[228] Even his date is uncertain; the usual dates given (e.g. by Heath and most of the reference books) are *c.* 408–355 BC, since, according to Apollodorus (who in the second century BC wrote among other things a work on chronology), Eudoxus' *floruit* (usually taken to be a person's fortieth year) was in the 103rd Olympiad (i.e. 368–365 BC) and he was fifty-three years old when he died. On the other hand, according to Pliny (*Nat. Hist.* xxx, 3), Eudoxus stated that Zoroaster (a semi-mythical king of Babylon, who was also connected with the teachings of the Magi) had lived six thousand years before the death of Plato (348) – which, therefore, must have occurred in Eudoxus' lifetime. Gisinger[229] accepts this and dates him to *c.* 395–342, while De Santillana[230] prefers 390–337; thus both these datings are consistent with Eudoxus' death at the age of fifty-three, but involve the rejection of Apollodorus' other statement as to his *floruit* (unless indeed a man's *floruit* can be taken as occurring in his late twenties).

From the biographical notice in Diogenes Laertius (viii, 86–91) we learn that Eudoxus studied under the mathematician Archytas (see p. 76), and came to Athens at the age of twenty-three where he heard Plato and the sophists. Later he went to Egypt[231] and stayed sixteen months absorbing the astronomical knowledge of the priests of Heliopolis; according to Strabo (C 806–07) the site from which Eudoxus made his observations was still pointed out to visitors. Strabo also reports a story that Eudoxus accompanied Plato to Egypt where the two of them stayed for thirteen years, but this is certainly false and is merely another instance of the Greeks' fondness for manufacturing connections between their own sages and the 'wisdom of the orient'. Afterwards, Eudoxus apparently taught at Cyzicus (on the southern shore of the Propontis) and then returned to Athens (perhaps *c.* 350) with a number of his pupils; some time after Plato's death, it would seem, he went back to his native city, where he was received with civic honours and continued to write and study until the end of his life – Strabo (C 119) reports Poseidonius as mentioning Eudoxus' observation post at Cnidus from which he observed the star Canopus (α Carinae).

It was during his second stay in Athens that Eudoxus, now a recognized scholar, seems to have joined in philosophical discussions with members of the Academy including Aristotle; there are reports in our sources[232] of bad feeling between the school of Eudoxus and the Platonists, and certainly one of his philosophical beliefs, that Pleasure is the *summum bonum*,[233] would seem to be fundamentally opposed to Platonic thought. On the basis of such evidence, Frank[234] constructs an elaborate hypothesis of a Eudoxus imbued with a strongly empirical, scientific attitude derived from Pythagoreanism (Archytas was a Pythagorean) and opposed in his whole philosophy and outlook to the metaphysical and speculative attitude of Plato and the Academy. He is pictured as entering into formal debates with representatives of the Academy, particularly Speusippus (Plato's destined heir) – debates which were attended also by the youthful Aristotle, whose rejection of the doctrine of Ideas may have been influenced by Eudoxus. All this is highly conjectural, and even less convincing is Frank's suggestion[235] that Eudoxus and his school made an astronomical expedition to Syracuse for the express purpose of observing the solar eclipse of 361 BC! What does seem certain, however, is that there was ample opportunity for the exchange of ideas between Eudoxus and Plato and their respective schools, and that while the latter's increasing interest in astronomy in his old age may have been owing to his awareness of the lines on which Eudoxus was working, these themselves might well have been instigated by Plato's call for a more mathematical treatment of the subject.

In the history of astronomy Eudoxus is important as being the first Greek astronomer of whom we have definite evidence that he worked with and fully understood the concept of the celestial sphere, and the first to have attempted the construction of a mathematically based system that would explain the apparent irregularities in the motions of the sun, moon, and planets as seen from the earth. As we have noted (p. 75), there is little evidence that even the Pythagoreans, for all their insistence on the importance of number in the universe, went far in applying mathematical methods to the solution of astronomical problems; and theories concerned with celestial phenomena were expressed largely in qualitative terms with a noticeable dearth of numerical factors. It was Eudoxus, a mathematician of genius,[236] who first seems to have realized the necessity of quantifying observational data and treating them mathematically, and as such he may rightly be termed the founder of mathematical astronomy.

The two works in which he evidently described the celestial sphere, its chief circles, their relationships to each other, and the constellations that marked them, were entitled Φαινόμενα and Ἔνοπτρον (the *Mirror* – presumably as being a descriptive image of the heavens). Their contents are known solely through the quotations of them by Hipparchus in the course of his *Commentary*[237] on the poem of Aratus, itself called *Phainomena* (third century BC), which gives a popular, non-mathematical exposition of the chief features of the celestial globe, describes the main constellations, indicates what stars rise or set with the zodiacal constellations (to facilitate telling the time at night), and ends with a number of weather prognostications based on celestial and meteorological phenomena. The astronomy of the poem is based on the work of Eudoxus,[238] a fact which Hipparchus proves[239] by a comparison of individual passages (*Comm. in Arat.* i, 2), and the greater part of his *Commentary* is taken up by detailed criticisms of the astronomical knowledge (particularly

of the positions of the various constellations) displayed by
Eudoxus, Aratus, and other commentators on the poem. Such
criticism was not given in any captious spirit, but because
Hipparchus was concerned that the astronomical errors com-
mitted should not be perpetuated for want of authoritative
rebuttal. He is, in fact, at some pains to absolve Aratus from
blame, since he was merely following Eudoxus and was, anyway,
a poet and not an astronomer (i, 1, 6–8).

According to Hipparchus, the subject-matter of Eudoxus' two
treatises was the same 'except for a very few details', and Aratus
based his poem on the *Phainomena* (πρὸς τὰ Φαινόμενα δὲ τὴν
ποίησιν συντέταχεν, i, 2, 2). However these 'details' included
different relative placings of constellations (i, 6, 1–2; 8, 6–7), and
Hipparchus even records Aratus' evident puzzlement as to which
version to follow in such cases (ii, 3, 29–30), so that it would seem
that the poem made use of data derived from both the treatises.
It is possible that the one is a revised edition of the other, in which
case it seems likely that the *Enoptron* is the later work, for we are
told (i, 2, 22) that in it Eudoxus gave the ratio of the (summer)
tropic above the horizon to that below it as 5:3 (i.e. a longest day
of 15 hours, equivalent to a latitude of 41° north, as Hipparchus
himself says at i, 3, 7),[240] while in the other treatise (which, in
the context, i, 3, 10, can only be the *Phainomena*) the same ratio
is given as 12:7, equivalent to a latitude of 42°21′, over 1° further
north. Since, according to the admittedly conjectural details of
his life history, his teaching activities at Cyzicus (latitude 40°23′)
were apparently earlier than his work at Athens (lat. 38°) or
Cnidus (lat. 36°43′), then the *Phainomena* may have been compiled
first.[241]

Hipparchus takes Aratus and his commentators (who should
have known better) to task for supposing that either of the above
solstitial ratios corresponded to the latitude of Greece, i.e. Athens,
which Hipparchus gave as 37° (1° too low – see my *GFH*, p. 177
ad fin.). Unfortunately, it is not clear what regions Eudoxus
himself connected with the ratios, and it is even possible that they
represent two measurements for the same region, which might,
in fact, have been intended for Greece; there is only a difference

of 9 minutes in the lengths of the longest days corresponding to the ratios[242] and it is very doubtful whether such a difference would be perceptible by the methods of time measurement available to Eudoxus. Hipparchus, of course, working with latitudes measured in degrees, more accurate instruments, and a knowledge of trigonometry, noticed the discrepancy (just over $1\frac{1}{4}°$), but it is clear from his words at i, 3, 11–12 that he also realized the inadequacy to be expected of astronomical theory of a period some two hundred years earlier than his own, and was prepared to be satisfied if the data he was examining agreed only roughly with the latitude of Greece. The observer's latitude would anyway only make a marked difference as regards the risings and settings of constellations, while the general description of their relative positions on the celestial sphere would not be much affected by it, except in so far as more of the stars near the south pole would become invisible as the north latitude of the observer increased.

Hipparchus tells us that Eudoxus described the constellations through which the chief circles of the celestial sphere passed – the summer tropic (*Comm. in Arat.* i, 2, 18; 10, 13–15), the winter tropic (i, 2, 20; 10, 17), the equator (i, 10, 22–3), and the solstitial and equinoctial colures[243] (i, 11, 9–21; ii, 1, 21). In addition, he gave the stars which marked the 'arctic' and 'antarctic' circles for the region of Greece, i.e. the limits of the circumpolar stars and those never visible at that latitude (i, 11, 1–8). Hipparchus has many criticisms to make of Eudoxus' relative placings of the constellations, but also indicates agreement with a number of them;[244] he finds errors of up to 10° (although in the majority of instances the error ranges only from 1°–3°) in the descriptions of the tropics, equator, and 'arctic' and 'antarctic' circles, but up to 23° and once (astonishingly) of nearly two zodiacal signs, i.e. nearly 60° (i, 11, 16), in the description of the colures.[245] As regards the 'antarctic' circle, he finds fault only with Eudoxus' placing on it of the star Canopus which, Hipparchus says (i, 11, 7–8), is situated about $38\frac{1}{2}°$ from the south pole and so is visible at north latitude 37° (Athens) and 36° (Rhodes);[246] but he explicitly states that he has no fault to find with the other data that Eudoxus gives for this circle, which are near enough correct (τούτων δὲ τὰ μὲν ἄλλα, ἐπεὶ κατὰ συνεγγισμὸν εἴρηται, οὐκ ἂν διστάζοιτο). This agreement is particularly significant in view of some of the theories propounded by modern scholars concerning the origin of the constellations – see below. Hipparchus criticizes Eudoxus for saying that there is a star at the north pole, and points out that there is actually an empty space there, but that the hypothetical point marking the pole forms one corner of a quadrilateral with three other stars.247

We are not expressly told that Eudoxus listed the twelve zodiacal constellations marking the sun's annual path through the heavens, but it is certain that he did. This becomes clear not only from Aratus' description (525–58) which, as we are constantly told by Hipparchus, followed that of Eudoxus, but also from the fact that Hipparchus emphasizes that, whereas Aratus placed the tropical and equinoctial points (σημεῖα) at the beginning of their respective zodiacal constellations (Cancer and Capricorn for the tropics, Chelae – later better known as Libra – and Aries for the equinoxes), as did Hipparchus himself and most of the earlier Greek astronomers (ii, 1, 19), Eudoxus placed them in the middle (ii, 1, 15; 20); so that the latter's zodiac is, as it were, shifted back half a sign compared with the Aratean or Hipparchian zodiac, and e.g. the beginning of Cancer for Eudoxus was on the later system 15° Geminorum (ii, 2, 5; cf. i, 6, 4). However, knowledge of the zodiacal *constellations* (ζῴδια) does not necessarily entail knowledge of the zodiacal *signs* as twelve equal sectors of the ecliptic (δωδεκατημόρια). In the section of his *Commentary* that deals with risings and settings248 (ii, 1–3), Hipparchus criticizes Aratus and his commentators for supposing that, if the sun's position on the ecliptic is known, the hour of the night can readily be determined by observing the rising of a zodiacal constellation or (if weather conditions prevent this) the position of extra-zodiacal stars known to rise at the same time as it (ii, 1, 2f.). This, he points out, is based on the fallacious premises that (a) the zodiacal constellations all take the same time to rise, and (b) that they are identical with the zodiacal signs, whereas in fact some of the constellations are smaller than the respective signs, e.g. Cancer, and some larger, e.g. Virgo and Leo (ii, 1, 7–8). It is clear that this criticism applies to Eudoxus as well, for Hipparchus goes on to find many errors in the data on simultaneous risings given by Aratus and Eudoxus (although he also confirms much of it, and sometimes upholds the one against the other – e.g. ii, 2, 10). He frequently emphasizes that Aratus derived his information from Eudoxus,[249] differing only in his placing of the tropical and equinoctial points and in a few other details, and often couples them together as 'they' (e.g. ii, 2, 11, . . . ὁλοσχερῶς δοκοῦσί μοι ἀγνοεῖν· φασὶ γὰρ . . .).

It is evident that Eudoxus' understanding of the sun's course was still far from complete, since he supposed that its path could be plotted with reference to the visible, zodiacal constellations and that these follow each other over the horizon at equal intervals; this is consistent with the assumption (which we know he definitely made – see below on his system of concentric spheres) that the sun travels with uniform speed round the earth. Certainly, Eudoxus could not have known the zodiacal signs as segments of the ecliptic exactly 30° long, since he did not use the 360° division of the circle;[250] all the degree figures given in the *Commentary* are Hipparchus' own and none of them is a citation from Eudoxus or Aratus. Equally certainly, since he described the tropics, equator, zodiac, and colures by the constellations through which they passed, he must have worked with some figure for the obliquity of the ecliptic. The fact that the apparent path of the sun is inclined to the equator seems to have become known in the last half of the fifth century BC and, as we have seen, Oinopides, a younger contemporary of Anaxagoras, is reported to have discovered this. No actual value for the obliquity is attributed to Oinopides (Eratosthenes is the first to have made an accurate measurement according to our evidence), but it is possible that the rough estimate equivalent to 24°, expressed as the angle subtended at the centre of a circle by the side of a regular, inscribed, fifteen-sided figure, may have originated with him and so was probably known to Eudoxus. The construction of a fifteen-sided figure within a circle is given by Euclid iv, 16, and Proclus says (*Comm. in Eucl. i*, p. 269, 8ff. ed. Friedlein) that this was inserted because of its usefulness in astronomy – Hipparchus also uses this approximate value when criticizing Aratus' placings of the constellations (*Comm. in Arat.* i, 10, 2), although, of course, knowing a more accurate figure (*GFH* fr. 41).

As an illustration of the type of material we are dealing with here, the following may be quoted (i, 2, 18 – part of Hipparchus' demonstration that to a large extent Aratus' poem is a paraphrase of Eudoxus):

> As regards the stars that move on the summer and winter tropic and also on the equator, Eudoxus says as follows with respect to the summer tropic, 'There are on this the middle section of the Crab and the longitudinal part of the body of the Lion, a small part of the upper section of the Virgin, the neck of the gripped Serpent, the right hand of the Kneeler (i.e. Hercules), the head of Ophiuchus, the neck of the Bird (i.e. Cygnus, the Swan) and its left wing, the feet of the Horse (i.e. Pegasus) . . .', etc., etc.

Later on, in the course of his detailed criticism of these data, Hipparchus remarks (i, 10, 14–15),

> At any rate it is the head of Ophiuchus and not the shoulders that Eudoxus says lies on the tropic, a statement which is itself false; however, the head is nearer the tropic than the shoulders, for the former is almost exactly 7° south of the summer tropic . . . It is Eudoxus' own opinion that the neck of the Serpent, which Ophiuchus holds, and the right hand of the Kneeler lie on the tropic, which is also consistent with observation.

This type of verbal description, relating the particular star to its individual position in the shape of the constellation, remained the normal method of referring to a star throughout antiquity and the Middle Ages right up to the time of Bayer (beginning of the seventeenth century). It was he who first designated the stars in a constellation by letters of the Greek and, when he had used these up, Roman alphabets, and it is his letter system that is commonly used to this day.[251]

Now it is obvious that this early system of reference, if it is to be used with meaningful results, involves two presuppositions: it assumes (*a*) that the sky is mapped into constellation figures generally recognized and known by all observers, and (*b*) that such figures have been reproduced in pictorial form so that their details can be readily referred to and communicated to others. This second assumption implies that Eudoxus worked with either a celestial globe or some sort of planisphere, and the former seems much the more likely.[252] Not only would it be easier to represent the chief circles on a globe which, by its rotation, could then furnish directly by simple inspection further data concerning risings and settings (which would otherwise have to be calculated from the fixed positions on a planisphere), but we also learn from Ptolemy that his own star catalogue was compiled by comparison with the representations on the 'solid globe' (στερεά σφαῖρα) of Hipparchus (*Synt.* vii, 1, ed. Heiberg, vol. ii, p. 11, 22), and there seems no reason to doubt that the latter similarly compared the data on his own globe with those on an earlier Eudoxan model which, to judge from the extent of Hipparchus' criticisms, was not a very accurate one – as might well be expected.[253]

The first assumption raises the baffling question of the origin of the ancient constellation figures (some forty-eight in number) as they have come down to us; how and when did all those human, animal, and other figures come to be associated with particular groups of stars? The question is probably unanswerable. The stories told to account for the names of the constellations (in various sources such as the pseudo-Eratosthenic *Catasterismoi*, the *Astronomica* attributed to Hyginus, and Ovid's *Fasti*) provide no answer, since they are obviously mere mythological rationalizations of an already existing state of affairs. Webb, who has written more sensibly than most on this subject, about which a great deal of nonsense has been written, was firmly of the opinion that in the great majority of cases the constellations received their names because the stars forming them really did suggest to the ancient stargazers the likenesses of the various figures – the Kneeling Man (later Hercules), the Hunter with club and belt (Orion), the Man Holding the Serpent (Ophiuchus), the Bull with its well-defined horns, the Scorpion with its claws and tail, and so on.[254] Certainly, it is undeniable that different peoples inhabiting different parts of the earth (American Indians, Australian aborigines, African Bushmen, etc.) have invented names and stories (in several instances similar to those we know from the Greeks) connected with the same conspicuous groups of stars in the sky,[255] and that the names of some of these at least (e.g. the Northern Crown, the Arrow, the Triangle, and perhaps also Draco and Hydra) appear self-explanatory (but see below).

Other writers have supposed that the constellations came into being on some preconceived plan at a definite time and place, thought to be calculable from consideration of the extent of the vacant space left round the south pole of a celestial sphere when all but the forty-eight ancient constellations have been removed from its surface.[256] Thus Maunder [257] asserts that the constellations were 'designed' at *c.* 2,700 BC by a people living between 36° and 40° north latitude, while Allen [258] suggests 2,000–2,400 BC. Such estimates are, of course, highly unreliable; the vacant space is by no means circular and has been described as at best 'irregular sausage-shaped'; its boundaries cannot in any event be defined at all accurately because we do not now know the limits of the original forms of the constellations – for if one thing is certain it is that these limits have been altered by successive generations of astronomers from antiquity to the present day.[259]

Again, Maunder speaks of the constellations as being 'designed', so that Draco linked the north poles of the equator and the ecliptic, Hydra lay along the celestial equator, Serpens partly along the equator and then bent at right angles to mark the line of the equinoctial colure and the zenith with its head (!), and the foot of Ophiuchus pressed down on the Scorpion's head 'just where the colure, the equator, and the ecliptic intersected' (*op. cit.*, p. 159). Similarly, Ovenden[260] comes, although by somewhat different methods, to much the same conclusions as Maunder concerning the date and latitude of the hypothetical constellation-makers and uses the above illustrations in the same way. He is convinced that the constellations are 'orientated symmetrically with respect to the celestial poles of about 2,800 BC' (p. 6), that they were designed 'as a primitive form of celestial co-ordinates' (p. 8), and that their arrangement actually contains two systems (!), a zodiacal one for 'the telling of the seasons' for agricultural purposes (with the constellations arranged symmetrically with respect to the ecliptic pole) and an equatorial one for the purposes of navigation (with the constellations arranged symmetrically with respect to the celestial pole).

Now all this is entirely illusory and affords an excellent example of something to which I have drawn attention elsewhere,[261] 'the tacit assumptions, based on the scientific theory of late antiquity or even (sometimes) of our own times, that underlie so much of the writing about early Greek science – assumptions for which there is no evidence, but which are almost unconsciously made from our inability to dissociate our views on the thought of this early period from more modern concepts'. To suppose that certain constellations were designed to mark the equator, the ecliptic, and the equinoctial colure and their point of intersection, or that they were arranged on a definite plan in relation to the celestial and ecliptic poles, is to assume that the whole concept of the celestial sphere with its great circles (which are after all mathematical abstractions of a decidedly sophisticated kind) was already known and understood by these putative designers; one cannot have poles and an equator without a sphere, or an ecliptic and colures without a tolerably accurate idea of the sun's circular path round the sphere, both daily and annual.[262] There is not the slightest justification for such an assumption, and the evidence which we have of the historical development of astronomical concepts flatly contradicts it. Neither in Egyptian nor in Babylonian astronomy do we find the concept of the celestial sphere until the period after its development by the Greeks, and we have traced its slow evolution in Greek astronomical thought from the early Pre-Socratics to Eudoxus. The belief that such a comparatively advanced stage of astronomy was reached by Minoans in the third millennium (and then apparently vanished, only to be laboriously rediscovered millennia later) can, of course, be maintained. So can a belief in the lost civilization of Atlantis or the land of the Houyhnhnms, and with as much reason; but theories concerning the origin of the constellations that entail such a belief are obviously beyond the range of rational discussion.

Another entirely gratuitous assumption that is commonly made is that the data recorded for the Eudoxan-Aratean celestial globe refer not to the heavens of that time but to those of a much earlier epoch. This assumption is apparently based on the number of criticisms made by Hipparchus, and a fixed belief that Eudoxus or Aratus cannot possibly have made mistakes; if the text makes it clear that they did (by comparison with the correct data for their time and latitude), then another time and latitude must be found more nearly corresponding to the data. An indication of the extreme unreliability of such comparisons[263] is given by the widely differing estimates of the date considered appropriate for the Eudoxan-Aratean globe, which vary from 3,000 to 1,000 BC. Thus Allen (*op. cit.*, p. 17) informs us that the Aratean sphere '*accurately* represents the heavens of *c.* 2,000 to 2,200 BC' (my italics); Brown[264] says that the Aratean data '*exactly* agree with the actual state of things at the vernal equinox 2,084 BC, a date when the Euphratean formal scheme or chart of the heavens had been already completed' (my italics); Ovenden (*op. cit.*, p. 12) gives a date of 2,600 ± 800 years; while Böker (in a paper that must surely represent the high water mark in unhistorical, pseudo-scientific speculation in this field[265]) estimates with totally spurious accuracy the epoch as −1,000 ± 30–40 years and the latitude as between 32°30′ and 33°40′.

All such estimates are based on fallacious premises. There is no reason at all to assume that the Eudoxan data reproduced by Aratus are anything other than they purport to be, namely, an early (probably the earliest) attempt to describe the celestial sphere and its chief circles as they appeared in the fourth century BC. Naturally, many mistakes were made, and these Hipparchus, living some two hundred years later, and having a far better grasp of astronomical theory and practice than was possible in Eudoxus' time, noted and corrected in the interests of accuracy, since they were in danger of being perpetuated by ignorant commentators on Aratus' poem (*cf.* his own explanation of the purpose of his *Commentary* – i, 1, 5–7). Some of the smaller discrepancies he found may well be put down to the effect of precession,[266] as it is certain that he wrote his *Commentary* before he had discovered this phenomenon – see *GFH*, p. 17. It is absurd to suppose that, because some of the data are wrong for the time of Eudoxus (according to modern calculations) but happen to suit a much earlier epoch, then all of them must necessarily refer to this epoch. What seem to have been completely disregarded by modern commentators are the numerous occasions when Hipparchus specifically registers his agreement with the Eudoxan-Aratean data.[267] Of particular significance is the fact that Hipparchus has only one criticism to make of Eudoxus' delineation of the stars that mark the 'antarctic' or always invisible circle, accepting all his statements as approximately true except that concerning Canopus (where, as can be shown – below, note 246 on *Comm. in Arat.* i, 11, 6–8 – Eudoxus was actually more correct than Hipparchus).[268] If the Eudoxan sphere had really been valid only for a time long anterior to him, such agreements could hardly have been registered, and changes in the limits of the never visible stars would have been among the first to attract Hipparchus' attention. To suppose that he might have failed to notice them entails an unjustifiably low opinion of his competence as an astronomer.

One source which suggests itself for the origin of the constellations is Mesopotamia – particularly since the researches begun some eighty years ago by Assyriologists such as Epping, Strassmaier, Pinches, and Kugler,[269] and continued by Weidner, Schaumberger, Neugebauer, and Van der Waerden (to name but a few more recent scholars) have made it clear that of the pre-Greek civilizations of the Near East only the Babylonians[270] developed their astronomical knowledge to a level in any way comparable with the Greeks, and undoubtedly influenced later Greek astronomy. It seems certain that at least from the Old Babylonian period (say, 1,800 to 1,500 BC) the Babylonians recognized various constellations in the night sky and used these as reference points in the description of astrological omina.[271] Lists of stars and constellations connected with different months in the Babylonian calendar are also known from this period or a little later, and although these texts are not yet fully understood it is possible that they represent schematic attempts to differentiate the times of the heliacal risings of stars.[272]

The mulAPIN texts (written *c.* 700 BC, but based on older material) show a division of the sky into three zones or 'ways', twenty-three stars or constellations being assigned to the 'way' of Anu (a belt extending some 15° each side of the equator), thirty-three to that of Enlil, and fifteen to that of Ea (the last two zones lying respectively north and south of Anu).[273] The identities of many of these star-groups are very uncertain, as can be seen from the data in P. F. Gössmann's *Planetarium Babylonicum*.[274] Many names can mean variously a star, a constellation (or part of one), or even a planet. Thus mul UZA, meaning 'goat star', can designate the head of the goat (β Capricorni), or Capella, or a constellation identified as Lyra or Lyra + θ Herculis, or the planet Venus (*op. cit.*, p. 60). The latter is also DIL. BAD, but has at least twenty other names, e.g. according to the month in which it appeared and whether it was in the east or the west (pp. 35*f.*). Similarly, mul ŠUDUN, meaning 'yoke star', is equated with Boötes and particularly Arcturus, but in astrological texts it can refer to Jupiter, Mars, or Mercury (p. 210).

Notwithstanding these difficulties, some of the more certainly identified Babylonian constellations definitely appear to be named after the same objects as their Greek counterparts. For example, both Babylonians and Greeks saw the stars Castor and Pollux as Twins, although the Babylonians also knew two other smaller twins, ζλ Geminorum and either γε or ημ or γξ Geminorum or αγ Orionis – p. 101. Both peoples saw roughly the same stars as a Lion (the Babylonians distinguished two, the second one apparently including the southern part of our Ursa Maior – p. 66; 69), a Snake, a Raven, an Eagle, a Fish (our Southern Fish), a Scorpion,[275] and a Bull. Regulus was the King star to the Babylonians also.[276] It is tempting to conclude from these similarities that the Greeks took over the constellations from the Babylonians, and this in fact is exactly what Van der Waerden claims, at least as regards the twelve zodiacal constellations.[277] On the other hand, apart from the numerous instances where the identifications are uncertain, there are as many (if not more) differences as similarities between the Babylonian and Greek constellations. The Babylonians saw a Plough where the Greeks had a Triangle + γ Andromedae,[278] a Bow and Arrow in the region of Canis Maior, a Goat instead of Lyra, a Hound instead of Hercules, a Great Swallow instead of the south-western part of Pisces, and Greek Andromeda was divided among three Babylonian constellations.[279] There is, in fact, no real justification for supposing that the similarities are anything other than coincidences between two separately developed constellation systems; until further evidence comes to hand, the Babylonian seems undoubtedly to have been the older, but that is as much as can safely be said.

The whole question of the influence of Babylonian astronomy on Greek in its early stages is fraught with considerable difficulties, not the least of which are the restricted evidence available to us for reconstructing the history of Babylonian astronomy before the Seleucid era[280] and the large measure of disagreement between modern authorities on the interpretation of this evidence. Herodotus (c. 480–425 BC) tells us in a much-quoted passage (ii, 109) that while geometry came into Greece from Egypt, where it originated from the necessity of land-surveying after the annual Nile floods, on the other hand 'the Greeks learnt about the "*polos*" and gnomon and the twelve parts of the day from the Babylonians'. The meaning of πόλος in this context (where it obviously does not mean the north or south pole or, as sometimes in poetry, the whole vault of the heavens – *cf. LSJ s.v.*) is not at all certain. From the fact that the word also occurs in a fragment of Aristophanes apparently concerned with telling the time by the sun,[281] it is commonly conjectured that the reference here is to a form of sun-dial consisting of a hemispherical bowl with an upright pointer set in the middle.[282] Despite its almost universal acceptance by modern scholars, this conjecture has little to recommend it. An upright pointer set in a hemisphere could not be used to tell the time (except midday) any more accurately than by simply noting the sun's position in the sky, unless the interior surface were engraved with such a multitude of lines (representing the changes in the position of the shadow corresponding to the continually altering altitude and azimuth of the sun) as to make it impracticable.[283] Nor is there any other evidence that such an instrument was used in this manner before the Hellenistic period. Hellenistic and later hemispherical sun-dials are engraved with lines representing the hour shadows at the solstices and the equinoxes,[284] and this clearly presupposes knowledge of the celestial sphere with tropics and equator; but this concept (like that of a spherical earth and of latitude) was never developed by the Babylonians.[285]

It is therefore more likely that '*polos*' is merely another name for 'gnomon', i.e. an upright pointer casting a shadow on a flat surface. This, probably the oldest 'scientific' instrument, could be used for marking the turning points of the sun's annual course, i.e. the solstices (longest and shortest midday shadows during the year) and noon (shortest daily shadow). The mulAPIN texts include a list of shadow lengths for a gnomon at different hours of the day on four days in the year, namely, the equinoxes and solstices. The data are very inaccurate for the latitude of Babylon (particularly the equinoctial and winter solstitial figures), which is not surprising since the underlying assumption seems to be that the length of the shadow increases in arithmetical progression with the height of the sun (which is, of course, incorrect). Moreover, the results are set out according to a predetermined scheme whereby the solstices and equinoxes are placed arbitrarily on the fifteenth day of the first, fourth, seventh, and tenth months of a schematic year of twelve months of thirty days each; probably the only actual observation was of the shadows at the summer solstice (which are less inaccurate than the others), and the rest of the data were calculated to fit the theoretical scheme.[286]

Herodotus, then, may well have been right in saying that the Greeks learnt about the gnomon from the Babylonians. He is less correct concerning the division of the day into twelve parts, since the fundamental Babylonian units of time were the 'double hour', *bêru*, the UŠ (30 UŠ = 1 *bêru*), and the GAR (60 GAR = 1 UŠ = 4 minutes), and it seems more likely that the twelvefold division of day and of night came to Greece from Egypt where it arose in connection with the use of the 'decans' as star calendars.[287]

The only other direct evidence for the influence of early Babylonian astronomy on Greek is Plato's remark that planetary observations reached Greece first from Egypt and Syria (*Epin.* 986e–987a; see above, p. 146), and Aristotle's statement that observations over very many years of the occultations of stars by planets had been made by the Egyptians and Babylonians 'from whom we possess much sound knowledge concerning each of the planets' (*De Caelo* 292a; *cf. Meteor.* 343b – although, as we have seen (p. 146), the reference to the Egyptian contribution is not borne out by our knowledge of the actual state of Egyptian astronomy).[288] An indirect allusion to the Babylonian origin of Eudoxan data is perhaps provided by Simplicius; in reporting the sidereal (zodiacal) periods which Eudoxus used in his planetary system (on which see below), Simplicius mentions the 30-year period for the planet Saturn, 'which,' he adds, 'the ancients used to call the star of the sun' (*in De Caelo*, p. 495, 28–9 ed. Heiberg, ὃν ἡλίου ἀστέρα οἱ παλαιοὶ προσηγόρευον). This same appellation is mentioned by several other Greek sources,[289] and Diodorus states that it was the Chaldeans (i.e. Babylonians) who called Saturn the star of the sun (ii, 30, 3), which is in fact attested by the cuneiform texts. It therefore appears that Simplicius' οἱ παλαιοί refers to the Babylonian astronomers, and it is possible that he added this gratuitous piece of information because the planetary periods used by Eudoxus were taken from Babylonian sources.

Now to achieve a knowledge of the periodic cycles of planetary phenomena, observations over a long period of time are obviously necessary, and it is just these that are lacking in Greek astronomy where, as we have seen, there is no good evidence that the planets were even distinguished from the fixed stars before the latter half of the fifth century BC.[290] On the other hand, we know that the Babylonians made observations of Venus in the middle of the second millennium BC:[291] the mulAPIN texts (which seem to summarize the state of Babylonian astronomical knowledge about 700 BC) contain some planetary data, and there exist other texts which include planetary observations, the earliest of which date[292] from −363 for Mercury, from −586 for Venus, from −422 for Mars, from −525 for Jupiter, and from −422 for Saturn, together with numerous undated texts. Unfortunately, notwithstanding this centuries-old collection of material, there is no direct evidence from the cuneiform tablets themselves that the Babylonians knew any accurate planetary periods before the Seleucid era, and the mulAPIN texts contain not a single period for the sun, moon, or planets, apart from the schematic year of twelve months of thirty days each.[293] Further, the period between the mulAPIN texts and the Seleucid texts, i.e. c. 700–312 BC – just the time when (at least from 450 BC onwards) it might be expected that Greek astronomy would have profited most from Babylonian – is the most conjectural in the reconstruction of the development of Babylonian astronomy and has given rise to the most disagreement between modern authorities.

No attempt can be made here to discuss the evidence for this controversial period, but it is important to gain some idea of the issues involved and of the nature of Babylonian astronomy, in so far as this is relevant to the history of Greek astronomy. Five points deserve to be emphasized: (a) the format of the cuneiform texts themselves; (b) the different goals aimed at by the Babylonian and Greek astronomers; (c) the schematization of Babylonian astronomical methods; (d) the difficulty of distinguishing between data obtained by actual observation and by calculation in Babylonian astronomy; and (e) the difficulties involved in the transmission of Babylonian results and methods to the Greeks.

(a) The texts consist for the most part of bald statements of astronomical data (positions of the moon and planets relative to certain stars at certain dates, times of risings and settings, eclipses, even meteorological phenomena such as haloes round the moon or sun, etc.), often in the early period connected with astrological prognostications for the king and his realms, or simply of lists of numbers in columns (especially in the Seleucid period), including dates, and referring to positions and times of lunar or planetary appearances. These lists require skilful decoding before one can understand the principles on which they were constructed, and there are relatively few texts which give any hints as to the procedure followed (except for occasional instructions such as 'multiply this by that' or 'add this to that and you see . . .'). Hence, although much is now known about Babylonian lunar and planetary methods,[294] there still remain earlier texts (and even parts of those belonging to the Seleucid era)[295] the meaning of which is uncertain. There are no continuous texts of descriptive astronomy or any discussion of astronomical principles such as we find in Greek; it is rather as though the star catalogue in the latter half of Book vii and the first half of Book viii and the solar, lunar, and planetary tables alone had survived of Ptolemy's *Almagest*.

(b) The chief aim of the Babylonian astronomers was to discover periodic relationships for the various celestial phenomena in which they were interested, to enable these to be predicted as accurately as possible (i) for the purposes of omen astrology and (ii) for the improvement of the Babylonian calendar, which remained at all periods a strictly lunar one, each month beginning with the appearance of the lunar crescent. The phenomena in question are mainly horizon phenomena (except for eclipses and the 'stations' of the planets), i.e. first or last appearances on the east or west horizon and periods of visibility of the moon, the stars, and the planets; these are what might be termed the 'natural' phenomena which an observer would notice, but there is no evidence that the Babylonians ever progressed to the stage of considering the celestial motions in their entirety. There is no trace in Babylonian astronomy (as at present known) of any desire to construct a comprehensive, mathematically based scheme which would account for all astronomical phenomena, or to investigate their nature and causes, or to take a synoptic view of the universe as a whole. This is in great contrast to the Greek thinkers, who from Pre-Socratic times onwards concerned themselves with just these matters; and, as Neugebauer points out, the different emphasis implicit in the approach of the later Greek astronomers to planetary motion as a whole 'marks an enormous step forward'.[296] Thus the Babylonians never developed the concept of the celestial sphere or of terrestrial latitude and longitude, and hence many of the theoretical problems of establishing the correct relationships between the basic circles of the sphere (equator, ecliptic, horizon, and colures), which form a major part of Greek astronomy, never appear in Babylonian astronomy at all.[297]

(c) The results and methods of Babylonian astronomy are set out in highly schematic fashion, the overriding concern being convenience of numerical manipulation and the desire to solve complicated periodic relationships by the use of successive approximations based on arithmetical progressions. One example of the schematization of data has already been given, namely, the 'gnomon tablet' of the mulAPIN texts (p. 166); the treatment of lunar and planetary data provides many other examples which can best be studied in the works of Neugebauer and Van der Waerden.

(d) The great emphasis placed on schematization often makes it difficult to decide whether the data in a particular Babylonian text have been actually observed or merely computed.[298] Neugebauer draws attention to ' . . . the minute role played by direct observation in the computation of the ephemerides. The real foundation of the theory is (a) relations between periods, obtainable from mere counting, and (b) some fixed arithmetical schemes (for corrections dependent on the zodiac) whose [sic] empirical and theoretical foundations to a large extent escape us but which are considered to be given and not to be interfered with by intermediate observations).'[299] For example, the dates of all solstices and equinoxes found in Seleucid texts have been shown to be the result of computation according to a fixed scheme, and not of observation,[300] so that no comparison of these with modern calculated dates is legitimate, and all attempts by such means to credit the Babylonians with the discovery of the precession of the equinoxes are invalid.[301] Moreover the lengths of the astronomical seasons were always assumed in Babylonian astronomy to be equal.[302] Similarly it has been shown that in the planetary theory all entrances of planets into zodiacal signs were computed and not observed.[303] On the other hand, *some* observations there obviously were, and those of eclipses, anyway, extended over an impressively long period; Ptolemy tells us (*Synt.* iii, 7) that Babylonian observations were available from

the beginning of the reign of Nabonassar, 747 BC, and he actually
uses in his calculations a lunar eclipse observed in Babylon in
721 BC (*Synt*. iv, 6).[304] However, he complains (as did Hipparchus
before him) of the lack of accurate observations of the fixed stars
before his time (*Synt*. viii, 1) and of the planets (ix, 2 – where he
emphasizes the difficulty of observing just those phenomena the
Babylonians were interested in), and had to rely on earlier Greek
material (e.g. *Synt*. x, 4, a Venus observation by Timocharis in
272 BC).

(*e*) Finally, the difficulties inherent in the utilization of Babylonian
material by a Greek astronomer must not be underestimated.[305]
Not only would he have to be able to read cuneiform writing
(or have access to accurate translations of the original), but he
would need a detailed insight into the completely different
methods and goals of Babylonian astronomy in order to extract
the results he required from the texts, which are not of such a
nature as to reveal the data on which they are constructed (e.g.
planetary periods) by simple inspection. Presumably this would
necessitate careful instruction by Babylonian astronomers, but
we have no evidence at all on these points, and *prima facie* it
would seem unlikely that such a situation should exist before the
opening up of the Near East to Greek influence after the conquests
of Alexander (died 323 BC).[306]

In view of all these uncertainties, it is perhaps not surprising
that modern scholars differ considerably in their interpretation
of the available evidence. Neugebauer paints a picture of the
relative independence of Greek astronomy from its Babylonian
counterpart in the early stages, and says that in contrast to the
situation in mathematics, where the Greeks had been able to
draw on a large stock of mathematical knowledge common to
the ancient Near East since Old Babylonian times, in astronomy
they had to start from scratch because (*a*) Babylonian mathematical
astronomy itself did not even begin to develop until the Persian
period (*c*. 540–330 BC), and (*b*) its aims and methods are very
different from those of Greek astronomy, particularly in its total
neglect of spherical geometry. He buttresses his views by pointing
out the marked difference between the high level of sophistication

of Greek mathematical treatises and the comparative crudity of astronomical theory until the second century BC – e.g. between Euclid's *Elements* and his *Phainomena*.[307]

One essential prerequisite for mathematical astronomy is a fixed system of dating and a regular scheme of intercalation; this was provided in Greek astronomy by the 19-year Metonic cycle introduced *c.* 430 BC (see above, p. 87). The same cycle is used in Babylonian astronomy, but its regular use is not attested before *c.* 380 BC, the texts before this time showing many instances of irregular intercalation.[308] Thus it is by no means impossible that the cycle was discovered by the Greeks before the Babylonians, who might have borrowed it from the former or discovered it independently. Again, in Neugebauer's view, it is impossible in the present state of the evidence to trace the early history of the Babylonian lunar and planetary schemes of the Seleucid texts or to give any dates for their invention;[309] nor do we know how far these were developed in the fifth and early-fourth centuries BC when they might have come to the notice of the Greek astronomers. Even the invention of the zodiac, which Neugebauer in common with most modern scholars takes to be of Babylonian origin, need not necessarily be so. The earliest Babylonian text that appears to mention zodiacal signs is a horoscope dated to 410 BC by Sachs;[310] for although the mulAPIN texts exhibit the arrangement of the three 'ways' of the heavens (see above), in each of which (twice in Anu) the sun stands for three months of the year, these 'ways' are parallel to the equator and defined by constellations, and there is no hint of the recognition of the sun's slanting path along a belt marked by equally spaced signs.[311] Now, as we have seen (p. 88), the invention of the ecliptic is attributed by Greek sources to Oinopides in the latter part of the fifth century BC, and it seems clear from what we know of the activities of Meton and Euctemon (p. 87) that they were also familiar with this concept. Hence it may well be that, in the case of the zodiac too, we ought to think of a parallel development in both Greek and Babylonian astronomy at about the same time, and not necessarily of any direct borrowing of the one from the other at this early stage.

A very different picture is drawn by Van der Waerden,[312] who is much more sanguine than Neugebauer about our knowledge of the early history of Babylonian astronomy, and believes that the evidence shows that a regular scheme of intercalation, first an 8-year cycle and then the 19-year cycle, was in use from 528 BC onwards with only two or three years as exceptions. He claims that it is possible to date the origins of the two systems A and B used in the lunar texts of the Seleucid era to the period −620 to −440, system A originating *c.* −500 and B some 40 years later, and that the corresponding planetary methods must have been invented between −520 and −300 and between −480 and −240 respectively.[313]

Much of the argumentation used to establish these dates is highly speculative (as Van der Waerden himself admits) and some of the conclusions drawn are extremely misleading. For example, he illustrates the 'ways' of the mulAPIN texts by a diagram showing a sphere with two circles parallel to the equator and a third intersecting these at an angle and marked by the twelve months of the schematic year (p. 78); the last circle, he states, is the zodiac, and therefore the Babylonians knew the oblique path of the sun, moon, and planets. This, however, is pure assumption. We describe the 'ways' as parallel to the equator, not because this is stated in so many words in the actual texts (which, of course, have no mention of 'equator'), but because of the positions of the identifiable constellations used to define them. The sun is stated to remain three months in each zone (six months altogether in Anu), but there is not the slightest necessity to assume that this must refer to a sphere with the sun's path oblique to the equator: the information in the texts could just as well be illustrated by a diagram showing three straight lines intersected by a perpendicular representing the sun's path. In fact, Van der Waerden's gratuitous introduction of the concept of a sphere is yet another example of the 'tacit assumptions' so often made without a shred of supporting evidence by modern commentators on ancient science;[314] because he finds it impossible to envisage the sun's path through zones in the sky without the familiar concept of the celestial sphere, he automatically assumes that the

Babylonian astronomers must have felt the same, and this actually leads him to the assertion that the mulAPIN texts show that they knew this concept (p. 134)!

Another instance where Van der Waerden draws conclusions that are totally at variance with the available evidence is in his discussion of Thales' alleged prediction of a solar eclipse (pp. 121f.). In a desperate attempt to vindicate the historicity of this prediction,[315] he spins a web of inferential reasoning, based on wholly improbable suppositions, which forces him to assume that the Babylonians and Thales not only knew a 47-month eclipse cycle (for which there is not the slightest evidence), but also were aware of the moon's movement in latitude and recognized that in 47 months the moon returns 51 times to the same node again (47 months = approx. 51 draconitic months). This presupposes not only accurate observations, but also the concept of the ecliptic as a mathematical line from which the moon's apparent path deviates both north and south (the nodes being the intersection points of the two), and the assumption that such comparatively advanced astronomical knowledge was possible in the sixth century BC is ludicrous; as we have seen, all the indications are (on both the Greek and Babylonian sides) that such a stage was not reached until at least 150 years later.

Van der Waerden's readiness to believe that astronomy attained a high level so early, and his failure to understand the implications of such a belief, makes it easy for him to accept unquestioningly all late testimony attributing advanced astronomical knowledge to the earlier Pre-Socratics; thus he accepts Diogenes Laertius' statement that Anaximander set up a gnomon in Sparta to measure the solstices and equinoxes (on the extreme improbability of this, see above p. 45), and has no hesitation in attributing sophisticated astronomical concepts to the Pythagoreans and even to Pythagoras himself (whose semi-mythical travels in the Near East he naturally accepts as undoubted fact, and as providing the means whereby he introduced Babylonian ideas to the Greeks).[316] All this enables him to develop the view that the Greeks borrowed from the Babylonians at all stages of the development of astronomy from the sixth century BC onwards. However, the credibility of this view depends entirely on the early dating of the methods of Babylonian mathematical astronomy (contradicted, as we have seen, by Neugebauer and resting on flimsy evidence), and is considerably diminished by Van der Waerden's uncritical treatment of the Greek sources.[317] The possibility that astronomical knowledge developed independently in accordance with the different aims of the Babylonian and Greek astronomers does not seem to occur to him; but a dispassionate and critical analysis of the evidence can hardly fail to confirm that this was very largely the case, and that not until the time of Eudoxus was there any fruitful interaction between Babylonian and Greek astronomy.

How this interaction took place is not known, but the position of Eudoxus' home town Cnidus (which came under Persian domination in the sixth century and later formed part of the Delian League) is consistent with his having access to Babylonian astronomical knowledge. It is possible that his placing of the solstitial and equinoctial points in the middle of the signs (see above, p. 156) and perhaps the lengths he assigned to the astronomical seasons[318] were influenced by Babylonian practice, according to which the seasons were assumed to be of equal lengths, and the solstices and equinoxes were arbitrarily placed at day 15 of months I, IV, VII, and X in the theoretical year of 12 months of 30 days each.[319] As already mentioned (p. 167), he may have used Babylonian data for the planetary periods assumed in his system, but it is certain that he could have derived no help from this quarter for the general arrangement of the planetary orbits. The Babylonian order for the planets was Jupiter, Venus, Saturn, Mercury, Mars, or Jupiter, Venus, Mercury, Saturn, Mars, which bears no relation to their order in spatial depth. The Greek order of Venus, Mercury, Mars, Jupiter, Saturn does, but this is not found in Babylonian astronomy.[320] Nor could this source have provided him with the idea of concentric spheres, which again was completely foreign to Babylonian astronomical methods and is entirely Greek in conception.

Eudoxus' planetary scheme is a brilliant piece of geometrical reasoning, although it was destined to be superseded by the epicycles and eccentrics of later Greek astronomy (but not for 150 years). Its historical importance lies not so much in its being a great improvement on the Pythagorean Philolaic scheme nor in its acceptance by Aristotle (with modifications – see below), as in the fact that it is the first fully worked-out application of the geometry of the circle and the sphere to astronomical problems, which was to become such a powerful mathematical tool in the hands of Hipparchus and Ptolemy.

It is often stated that the Greek insistence on circular orbits for the heavenly bodies was a philosophical dogma which acted as a retarding influence on the development of astronomy.[321] This is at best only a half truth. Granted that circular motion, considered as especially apposite for the sphere as the perfect figure, and hence for the heavenly bodies, *did* become accepted philosophical doctrine, yet this was in the first place a wholly legitimate inference from the results of observation. The stars *are* seen to move in circular orbits across the sky, the sun *does* appear to go round the earth in a circle, and the assumption of a spherical shape for the moon and the earth (and therefore not unreasonably for all the celestial bodies) proved to be the only satisfactory way of explaining the relevant phenomena. What is commonly overlooked is the mathematical utility and convenience of operation of the concept of circular motion: reduce your observed periodic movements to circles and combinations of circles and at once you make them amenable to calculation and predictable as to both spatial position and time. The mathematics of the circle and sphere ('sphaeric') early became a special field for investigation by the Greek mathematicians, no doubt because of its usefulness in treating astronomical problems.[322] In fact, just as the chief mathematical tool of the Babylonian astronomers was arithmetic progressions,[323] so that of the Greek astronomers was circular motion, and the latter can no more be regarded as a hampering dogma than the former. It is one of Eudoxus' chief claims to fame that he seems to have been the first Greek astronomer to exploit the mathematical properties of the circle and the sphere in explaining celestial movements.

The Eudoxan system (described in one of his lost works

Περὶ ταχῶν, *On Speeds*) is reported briefly by Aristotle (*Met.* Λ 8, 1073b17ff.) and in more detail by Simplicius (*Comment. in Arist. De Caelo* ii, 12, 221a, pp. 493ff. ed. Heiberg). In modern times little note was taken of it until the nineteenth century,[324] when Ideler, Apelt, Martin, and Schiaparelli explained its main principles.[325] The last-named in particular demonstrated the mathematical elegance of the system, and showed that the geometry involved was well within the competence of a mathematician of Eudoxus' calibre. Schiaparelli's reconstruction (which is, of course, hypothetical, since the geometrical details are nowhere given in the ancient sources) has been generally accepted, and the accounts of Dreyer and Heath are both closely based on it.[326]

Eudoxus' system is a logical but sophisticated development of the early notion (going back at least to Parmenides and probably earlier – see also p. 51) of a spherical cosmos, and hence of the circular motions of the heavenly bodies (emphasized, e.g., by Empedocles – see pp. 53-4), combined with the idea of the earth as the fixed centre round which the orbits take place. The stages of this development are these. First, the orbits, including that of the fixed stars, were envisaged as being all in the same plane but completed in different periods, i.e., the fixed stars, sun, moon, and planets all travelled at different speeds round the earth. This stage is exemplified by the Philolaic system and the early Platonic 'spindle of Necessity' (see above). Secondly, it was recognized that the plane of the *annual* motions of sun, moon, and planets (each travelling at its own particular velocity in a kind of composite path, the band of the zodiac) must be inclined to that of the *daily* rotation of the whole celestial sphere, but that the latter rotation exerted an overriding influence on all the celestial orbits to account for the daily phenomena of rising and setting. This is the later Platonic picture as described in the *Timaeus*. Both these stages are characterized by uniform circular motion round a *fixed* point and (in the second stage) round *fixed* axes. Finally, in order to try to account for the anomalous movements of the planets in both longitude and latitude as evidenced by an increasing body of observational material, Eudoxus hit on the idea of investigating the composite motion of a body resulting from its uniform rotation round an axis which is *not fixed*, but the poles of which are themselves attached to an enveloping sphere which also rotates at a uniform speed round an axis different from the first one. Then, as proved necessary, the poles of this sphere could be envisaged as carried by yet another and so on. Thus Eudoxus substitutes concentric, rotating spheres (or rather spherical shells) for the plane circles of earlier astronomy, and the paths of moon, sun, and planets as seen from the earth are the results of the component motions of the various spheres, each of which exhibits uniform rotation. Each celestial body has its own individual arrangement of spheres, the rotational speeds and the axial inclinations of which are governed by the observed periods and deviations in latitude of the particular body, which in every case is supposed to be situated on the equator of the innermost (nearest the central earth) for each set.

For the moon, Eudoxus envisaged three spheres. The outermost of these rotated once in twenty-four hours round the poles of the cosmos from east to west and represented the diurnal movement of the whole heavens – this function of the outermost sphere was exactly the same in each of the seven sets of spheres for the seven planetary bodies. The second sphere had its poles fixed to the underside of the enveloping sphere, and, as well as being carried round by the latter, itself rotated round an axis perpendicular to the plane of the circle through the middle of the zodiac (so that, in effect, the equator of this sphere was the ecliptic) from west to east, i.e. in the opposite direction to the first sphere. Again, the function of this second sphere, namely, to represent the longitudinal movement in the zodiac in the opposite direction to the daily movement, was exactly the same in each set of spheres. The third sphere (which actually carried the moon on its equator), as well as being affected by the rotations of the other two spheres, since its poles were attached to the underside of the second sphere,[327] itself rotated round an axis inclined at an angle to that of the second sphere in the opposite direction to it and at a much slower speed. Simplicius tells us that the purpose of the third sphere was to account for the moon's movement in latitude (i.e. north and south of the ecliptic) and for the fact that its greatest deviations do not occur always at the same zodiacal points, but at points which shift steadily westwards; hence the inclination of the axis of this sphere to that of the second was fixed at the maximum amount of deviation.[328] From this it is clear (as is also implicit in Aristotle's account) that Eudoxus recognized only one anomalous motion of the moon, viz. its deviation in latitude, and otherwise regarded it as moving uniformly in the plane of the zodiac represented by the motion of the second sphere.

It is not known what speeds of rotation he assigned to the second and third spheres or what inclination he assumed for the axis of the latter; Simplicius merely says that the rotation of the third sphere was 'slow' and that the change in position of the points where the lunar latitude was greatest 'was extremely small each month'.[329] Since Eudoxus described the main circles of the celestial sphere in his *Phainomena* and *Mirror* (see above) by the constellations through which they passed, there is no reason to doubt that he could have noted approximately the moon's maximum deviation in latitude (which amounts to some $5°$, being the actual inclination of the mean lunar orbit to the ecliptic),[330] and this would give him the inclination of the third sphere; that of the second to the first would, of course, be the figure he assumed for the obliquity of the ecliptic (probably $\frac{1}{15}$ of a circle, equivalent to $24°$ – see above, pp. 157–58). The period of rotation of the second sphere would presumably be the lunar month[331] (already known with some accuracy by Euctemon and Meton), and the slow rotation of the third sphere in the opposite direction would not affect the final lunar velocity very much,[332] but would serve to explain why the maximum deviations north and south did not always occur at the same points of the zodiac.

The above reconstruction is based strictly on the ancient evidence. It reveals a still very imperfect understanding of the intricacies of the moon's movements, but this is no more than we should expect for the time of Eudoxus when the celestial sphere was a relatively new concept, and the only component of lunar motion that was likely to be well known was the period of its phases, i.e. the lunar or synodic month of about $29\frac{1}{2}$ days. Unfortunately, the picture has been confusingly blurred by the obstinate insistence of modern commentators on attributing to Eudoxus much more advanced astronomical knowledge than there is any evidence that he possessed. Thus Schiaparelli, followed by Dreyer and Heath and the many editors who depend on them,[333] unhesitatingly takes Simplicius' explanation of the third sphere (quoted above) as referring to the regression of the lunar nodes. In assuming that Eudoxus must have been familiar with this phenomenon, these scholars suppose that he intended the rotation of one of the spheres to represent this westward move-ment of the nodes which completes a full circle in about $18\frac{1}{2}$ years.[334] This gratuitous assumption then compels them to assert that both Simplicius and Aristotle[335] made a mistake as to the roles of the second and third lunar spheres, which must be interchanged so as to make the third sphere rotate in a month and the second in $18\frac{1}{2}$ years – thus directly contradicting Sim-plicius' explicit statement. For, says Heath (p. 197), 'If it had been the third sphere which moved very slowly, as he says, the moon would only have passed through each node once in the course of 223 lunations, and would have been found for nine years north, and then for nine years south, of the ecliptic'; and, says Dreyer (p. 92), '*Obviously* Eudoxus must have taught that the innermost sphere (carrying the moon) revolved in 27 days [the draconitic or nodical month] from west to east round an axis inclined, at an angle equal to the greatest latitude of the moon, to the axis of the second sphere, which latter revolved along the zodiac in 223 lunations in a retrograde direction' (my italics).

This is yet another example of a misleading interpretation of early astronomical thought by attributing to it concepts which were only discovered much later (*cf.* p. 161). It is highly improbable that Eudoxus knew anything about the nodes or distinguished the draconitic from the synodic month, and certain that he had no inkling of the true reason for the regression;[336] as we shall see, even Hipparchus, two hundred years later, could not formulate a satisfactory theory for the moon's motion, which is, in fact, extremely complex, so that even today discrepancies are found between its observed and predicted positions (see above, pp. 24f.). We know from Hipparchus that Eudoxus' stellar placings were

often very erroneous and that the globe he used was far from
accurate (as might be expected – see above).[337] What likelihood
is there then that his lunar observations would be any more
accurate or lead him to anything but a very approximate realization of the moon's deviations from its longitudinal path through
the zodiac? The fact that on his theory the moon must move in
a manner that does not accord with accurate observations of its
position indicates merely that scientific astronomy was still very
much in its infancy at that time; it is certainly no reason to dis-
believe what Aristotle (who undoubtedly knew much more
about his contemporary Eudoxus than we do) and Simplicius tell
us.[338]

For the sun, Eudoxus also assumed three spheres: the first two
had the same functions as in the lunar system except, of course,
that the period of rotation of the second sphere was a year instead
of a month. The third sphere also resembled its counterpart in
the lunar system, in that it was intended to represent a supposed
deviation of the sun in latitude, but its axis of rotation was
inclined at a smaller angle to that of the second sphere than in
the case of the moon (Aristotle), and its speed of rotation was
much slower and in the same direction (i.e. from west to east) as
that of the second sphere (Simplicius). Again, Schiaparelli and
his followers make the totally unfounded assumption that
Simplicius is wrong about the speeds of rotation of the second
and third spheres, which should be interchanged; otherwise, their
argument runs, the sun would remain for a long time north or a
long time south of the ecliptic, which is contrary to observation.
Again, there is no good reason to accept this. It is obvious that
Eudoxus' understanding of the sun's motion was just as imperfect
as of the moon's; not only did he postulate a wholly imaginary
deviation of the sun from the ecliptic, but it is clear that he assumed
a uniform, longitudinal, solar motion,[339] thus apparently ignoring
the inequality of the astronomical seasons which had been noticed
some eighty years before by Euctemon and Meton.

Simplicius says that the idea of three movements for the sun
was also held by astronomers before Eudoxus,[340] and explains
that the third sphere was postulated 'because the sun does not

always rise at the same place at the summer and winter solstices'
(p. 493, 15f.). It is difficult to know what to make of the first part
of this; if Euctemon and Meton held similar ideas to those of
Eudoxus on the sun's motion, then how is it that they were
able to give approximate values for the different lengths of the
seasons which would imply recognition of the non-uniform
course of the sun? It is easier to understand how observations at
the solstices might have led to the false assumption of latitudinal
variations in the sun's path. At the solstices the sun's declination
is altering very slowly; the midday shadows of a gnomon and
the length of daylight remain practically the same for several
days. Hence the precise moment of the solstice (which may
occur at any time of day and night) is difficult to estimate by
observation alone – and to this must be added the drawbacks of
inaccurate instruments and a crude observational technique. If
the estimated time of one solstice happened to be at one extreme
of the range of possible days and that a year later at the other
extreme (unfavourable weather conditions might easily cause this),
then there could well be a noticeable difference in the place of
sunrise on the two occasions. Several such occurrences, both
when the sun was north and when it was south of the equator,
would give rise to the theory of its deviation in latitude.

Eudoxus might the more readily have accepted this, since,
in his system, all five planets and the moon exhibited (genuine)
movements in latitude, and there was no reason to suppose that
the sun alone was exempt from the tendency. Confirmation is
provided by a verbatim quotation by Hipparchus of a sentence in
Eudoxus' *Mirror* which runs, 'The sun also appears to exhibit a
difference in position at the solstices, but this is much less evident
and quite small' (Hipp., *Comm. in Arat.* i, 9, p. 88, 19 ed. Manit.).
Hipparchus, of course, denies this fictitious latitudinal movement,
pointing out that if it really existed the prediction of eclipses
would be much less accurate than in fact it was, but the notion
persisted in writers such as Pliny, Martianus Capella, and some
commentators on Aristotle; Theon of Smyrna gave a figure of $\frac{1}{2}°$
for the maximum deviation each side of the ecliptic.[341]

For the five planets Eudoxus had to face a further problem.

As well as their deviations in latitude, he now had to explain why during their periodic circuits of the sky they are seen to remain stationary relative to the stars for several days, then retrace their paths westwards (retrograde motion) for a certain time, and remain stationary again before resuming their direct (eastwards) motion along the zodiac. The assumption of uniform motion in longitude with periodical deviations in latitude, represented by the second and third spheres for moon and sun, was alone no longer adequate to account even approximately for planetary movements. Eudoxus' solution was to postulate an extra sphere, making four in all, for each planetary set, the poles of each sphere being attached to the one enveloping it in the manner already described. The first sphere (outermost) represented as before the daily revolution of the whole heavens. The second sphere, again as before, represented longitudinal motion in the plane of the ecliptic and completed one revolution in the sidereal or zodiacal period of the planet. The third sphere had its poles on the equator of the second (i.e. the ecliptic) and, like the second, rotated from west to east but this time in the synodic period of the planet. The poles of this sphere were different for each planet, but the same for Mercury and Venus (Aristotle). The fourth sphere (carrying the planet on its equator) rotated round an axis inclined to that of the third (the angle of inclination varying with each planet) and in the same period as it but in the opposite direction.

The function of the third and fourth spheres was to account for the stationary points and retrograde movements of the planets and their variations in latitude. What Eudoxus sought was a mechanism whereby the planet's speed would be sometimes accelerated relative to its motion along the ecliptic (represented by the second sphere), would sometimes exactly counterbalance this (so that the planet would appear stationary in the sky), and would sometimes exceed it in the opposite direction (to give the appearance of retrograde motion). At the same time he wanted the planetary path to exhibit deviations in latitude. He found such a mechanism in the arrangement of the third and fourth spheres. Had he made the fourth rotate at the same speed in the opposite direction to the third and in the same plane, the motions would simply have cancelled out, with the result that the planet would have exhibited only the motions of the second and first spheres. Conversely, had he given no rotatory motion to the fourth sphere, but imagined the planet on its equator inclined at an angle to the third, the rotation of the latter would have caused far too great deviations in latitude (in fact, the planet would actually have been carried through the poles of the ecliptic). By making the fourth sphere rotate at an equal and opposite speed to the third *and* round an axis inclined to that of the latter, Eudoxus was able to confine the latitudinal movement of the planet within acceptable limits and also simulate its retrograde motions and stationary points; for it can be shown[342] that the resultant path of the planet is an oscillatory motion (with a period equal to the synodic period of the planet) round a track shaped like a figure-of-eight lying on its side (see Fig. 11), known as a lemniscate.[343] According to Simplicius, Eudoxus named it a 'hippopede' (ἱπποπέδη, literally 'horse-fetter'), a term also used by Proclus[344] to designate a similar figure. The 'hippopede' is itself carried round the zodiac (its longitudinal axis being the line of the ecliptic) by the motion of the second sphere rotating with the sidereal period of the planet, and the effect is that for half the synodic period the motion in longitude is accelerated by the motion along the 'hippopede' while for the other half it is retarded, the greatest acceleration and the greatest rate of retardation occurring at the double point (c, g) of the lemniscate. Therefore, when the speed of retardation exceeds the speed at which the whole lemniscate is being carried along the ecliptic, the planet appears to have a retrograde movement (seen from the earth), and when the speeds more or less balance each other (at a and e), the planet will appear to be stationary. As Heath explains (p. 207–08):

The movements must therefore be so combined that the planet is at the double point and moving in the forward direction at the time of superior conjunction with the sun, when the apparent speed of the planet in longitude is greatest, while it is again at the double point but moving in the back-

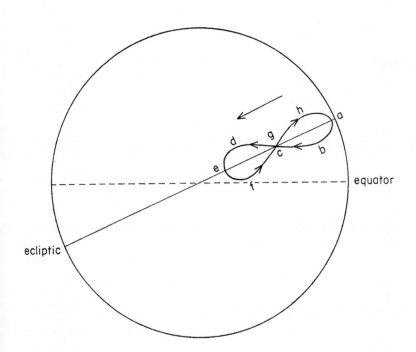

Fig. 11. The 'hippopede' of Eudoxus

ward direction when it is in opposition or inferior conjunction, at which times the apparent retrograde motion of the planet is quickest. This combination of motions will be accompanied by motion in latitude within limits defined by the breadth of the lemniscate; the planet will, during a synodic revolution, twice reach its greatest north or south latitude respectively and four times cross the ecliptic [i.e. at *a, c, g, e*].

Simplicius tells us the sidereal and synodic periods assumed by Eudoxus, and (except for Mars) these are roughly correct (see note for Mercury and Venus), as the following table shows:

	Sidereal period		Synodic period	
	Eudoxus	*modern*	*Eudoxus*	*modern*
Mercury	1 year	⎡ 88 days ⎤ [345]	110 days[346]	116 days
Venus	1 year	⎣ 225 days ⎦	19 months	
			(= 570 days)	584 days
Mars	2 years	1.88 years	8 months 20 days	
			(= 260 days)	780 days
Jupiter	12 years	11.86 years	13 months	
			(= 390 days)	399 days
Saturn	30 years	29.46 years	13 months	
			(= 390 days)	378 days

To reconstruct the Eudoxan system completely, we need to know also the inclination of the axis of the fourth sphere to that of the third for each planet; unfortunately, no ancient source gives us these inclinations. Schiaparelli (followed by Dreyer and Heath), taking the most favourable interpretation, i.e. assuming that Eudoxus knew the true lengths of the arcs of retrogradation,[347] shows that an inclination of 6° for Saturn would give the correct amount of retrogradation (about 6°), with a 'hippopede' of 12° in length (*a* to *e*) and a total breadth ($df = ha$) of 18′, producing a maximum deviation north and south of the ecliptic of 9′. For Jupiter, an inclination of 13°, with a 'hippopede' of 26° in length and a total breadth of 88′ would give the approximately correct retrogradation of 8°. These deviations in latitude (9′ and 44′), it is asserted, would not have been observable by Eudoxus' instruments (this is certainly true), and thus for these two planets the concentric sphere system corresponded excellently with the phenomena.[348] However, the very fact that such small deviations would have passed unnoticed throws considerable doubt on the legitimacy of attributing the above values to Eudoxus. One of the main features of his system was that it assumed deviations in latitude for *all* the planetary bodies and, as we have seen, this may well have been one of the reasons for the three spheres given to the sun. It would seem highly unlikely that he should have chosen such values as to produce deviations which were not perceptible. Schiaparelli's conjectural figures are no more than that, and probably bear little relation to the actual figures Eudoxus assumed.[349]

For Mars, the synodic period he accepted is three times too small; but, with an inclination of 34° and a 'hippopede' of length 68°, a retrograde arc of 16° and a maximum latitude of about 5° are produced, which are roughly right. Unfortunately, it can be shown that this entails two retrograde motions at parts of the orbit where the planet is not in opposition to the sun and four additional stationary points which are fictitious. If the true synodic period of 780 days is taken, the system will not work at all, since the planet will not become retrograde unless the inclination of the fourth sphere to the third is greater than 90°, i.e. unless the two spheres revolve in the same direction, which is contrary to Simplicius' description. Mars caused even Kepler a great deal of trouble;[350] Pliny refers to its *maxime inobservabilis cursus* (*Nat. Hist.* ii, 77), and it seems to be singled out for special mention in Plato's description of the 'spindle of Necessity' (see above, p. 112). For Mercury, the poles of whose third sphere are the same as for Venus (the centre of both 'hippopedes' being the sun), if the inclination of the fourth sphere is assumed equal to the greatest mean elongation (23°), then the maximum deviation in latitude becomes about $2\frac{1}{4}°$ (half the true figure) with a retrograde arc of 6° (much too small) in an unobservable part of its orbit. For Venus, on the same assumption of the inclination's being equal to the greatest elongation (47°), the maximum deviation in latitude is about 9° (correct), but (as with Mars) the planet can never have a retrograde motion no matter what value is assigned to the inclination. Whether Eudoxus in fact made the above assumptions is entirely uncertain.

We see, then, that the Eudoxan system of concentric spheres (three each for the sun and moon, four each for the five planets, and one for the fixed stars, making a total of twenty-seven) could only represent planetary motion in a very approximate fashion and had many drawbacks. According to it, the retrograde arcs for each individual planet are of the same length and occur at the same points of the ecliptic in every cycle. The deviations in latitude depend only on the planets' elongation from the sun, instead of on their longitude, and there is no knowledge of orbits inclined to the ecliptic – hence the planets are made to cross it too frequently. Since the second sphere in every set rotates uniformly in the sidereal period of the body, the system fails to account for zodiacal anomalies produced by the eccentricities of the orbits. This is especially striking in the cases of the sun and moon, where Eudoxus apparently ignored the inequality of the seasons and recognized only one anomaly for the moon, its deviation in latitude (see above). The theory postulates that each celestial body is at a fixed distance from the earth, and thus disregards the very noticeable differences in brightness at different times of, for example, Venus and Mars. It also disregards another easily observable phenomenon, namely, that Venus takes only 144 days to go from greatest eastern to greatest western elongation, but 440 days to go from greatest western to greatest eastern elongation. Nevertheless, with all its imperfections (only apparent from the hindsight afforded by later, more accurate observations), the system deserves an honoured place in the history of astronomy, and represents a considerable advance on the state of astronomical knowledge preceding it. It was no mean feat to evolve a geometrical explanation of retrograde movements and stationary points, and the application of mathematical principles to astronomical problems was an essential step in the progress of astronomy[351] – Dreyer is fully justified in saying (p. 107): 'Eudoxus is the first to go beyond mere philosophical reasoning about the construction of the universe . . . the science of astronomy has started on its career.'

Some of Eudoxus' astronomical work was concerned with calendaric problems. According to the Suda (s.v.) and Diogenes Laertius (viii, 87) he wrote a treatise entitled *Octaëteris*, but Achilles (*Isag.* 19, p. 47 ed. Maass) doubts the genuineness of this and says that Eratosthenes showed that it was not by Eudoxus – Censorinus mentions that the same title was attributed to several other writers (*De Die Nat.* 18, 5). An 'octaëteris' (ὀκταετηρίς) was, according to Geminus (*Isag.* 8, 27ff.), an intercalation cycle of 8 years containing 99 lunar months (three of which were intercalary) and 2,922 days, and this was the earliest cycle used by the Greeks to bring the solar year into correlation with the lunar month. However his account of its evolution is inconsistent, being based on a figure for the solar year of $365\frac{1}{4}$ days ($2,922 = 8 \times 365\frac{1}{4}$) which was not discovered until the time of Callippus, as Geminus himself tells us later. In fact, he goes to some lengths to demonstrate the inadequacy of the 8-year cycle, then describes the 19-year (Metonic) cycle as an improvement on it (this gives a year of $365\frac{5}{19}$ days – see above, p. 88), and finally the Callippic cycle of 76 years (giving a year of exactly $365\frac{1}{4}$ days) as a further improvement agreeing best of all with the phenomena.[352] Eudoxus may have ignored the 19-year cycle of Euctemon and Meton (as he apparently ignored their discovery of the inequality of the seasons – see above) and may have used and written about the 8-year cycle, but it can hardly have evolved as Geminus says it did. It is more probable, as Ginzel and Heath suggest (*Arist.*, p. 289), that it arose from the rough equation of 99 lunar months of $29\frac{1}{2}$ days ($= 2,920\frac{1}{2}$ days) with 8 solar years of 365 days ($= 2,920$ days). At any rate, it is certain that Eudoxus did some work in connection with a 'parapegma', since he is frequently cited as an authority for '*episemasiai*' (weather prognostications) in the calendar attached to Geminus' *Isagoge*, in Ptolemy's *Phaseis* (where it is stated that Eudoxus made observations 'in Asia and Sicily and Italy' – p. 67, 8 ed. Heib.), and in the other extant 'parapegma' texts.[353]

Eudoxus also seems to have initiated the application of mathematical principles to geography. Strabo (C 390) speaks of him as being familiar with the concept of 'climata';[354] this is certainly an exaggeration (see my *GFH*, pp. 160 and 24–5), but it is clear that Eudoxus understood how to define position on the earth globe by the relationship of the chief circles of the celestial sphere to the observer's horizon[355] (see above, pp. 154–55), and there is no reason to doubt that in geography, as well as in astronomy, his work marks the advent of the more scientific approach that was later developed by Eratosthenes, Hipparchus, and Ptolemy.

CHAPTER VII

CALLIPPUS AND ARISTOTLE

THE EUDOXAN SYSTEM of concentric spheres was further elaborated by Callippus of Cyzicus who was active as an astronomer in the latter part of the fourth century BC, and is frequently cited in the 'parapegma' texts.[356] According to Simplicius (*in De Caelo* ii, 12, p. 493, 5*f.* ed. Heib.), Callippus studied with a compatriot Polemarchus, who knew Eudoxus evidently during the latter's stay in Cyzicus (see above, p. 152). Callippus came to Athens after Polemarchus and stayed with Aristotle, correcting and amplifying Eudoxus' work. This Polemarchus is mentioned again by Simplicius as being aware of one of the drawbacks of the original Eudoxan scheme, namely, its failure to account for the variations in the distances of the planets as evidenced by their changing brightness. However he evidently regarded it as of little importance compared with the general plausibility of the scheme (*op. cit.*, p. 505, 21). It is clear that even during his lifetime Eudoxus' concentric spheres were recognized to be inadequate for the complete explanation of observed planetary phenomena,[357] but it was Callippus who made the most successful attempt to bring the theory more into line with observation.

Aristotle tells us (*Met.* Λ 8, 1073*b*32*ff.*) that Callippus kept the same arrangement as Eudoxus and assumed the same number of spheres for Jupiter and Saturn (four each), but that he thought it necessary to add two extra spheres each for the sun and moon, and one extra to each of the remaining planets, in order to account for the facts of observation;[358] so that altogether Callippus envisaged 33 planetary spheres instead of Eudoxus' 26. Aristotle's summary account gives no detailed reasons for Callippus' additions; according to Simplicius there was no book by Callippus extant to explain the additional spheres, 'but Eudemus related concisely the phenomena because of which he thought these spheres should be added'.[359]

However, the only specific reason Simplicius reports from Eudemus is the inequality of the astronomical seasons discovered by Euctemon and Meton (which he oddly connects also with the additional spheres for the moon, although it can have nothing to do with this); and as regards the planets he simply states again that Eudemus explained 'concisely and clearly' why the extra spheres for Mars, Venus and Mercury were added, but does not actually specify the reason. Schiaparelli, in his classic paper on the system of homocentric spheres (see above, p. 177), has demonstrated how Callippus' modifications could have improved the original theory. We know from the *Ars Eudoxi* that his values for the astronomical seasons were 92, 89, 90 and 94 days respectively, starting from the summer solstice;[360] he was therefore aware of the non-uniformity of the sun's course round the earth. If the two additional spheres for the sun were used in the same way as the last two spheres for a planet in the original Eudoxan system, then a hippopede would be produced which would be carried round the ecliptic and would have the effect of accelerating the sun's motion at certain parts of its course and retarding it at others. This was exactly the result required, and it can be shown that the sun's motion may be represented by this means almost as accurately as on the later epicycle or eccentric hypothesis. Similarly for the moon, the addition of two spheres to the original three of Eudoxus could be arranged to produce a hippopede which would account for the inequality of the moon's longitudinal movement, which is much more marked than that of the sun.

As regards the extra sphere each for Mars, Venus, and Mercury, we are not told why Callippus thought these necessary, but it seems highly probable that, for Mars and Venus anyway, the fifth sphere was designed to produce retrograde motions for these planets in accordance with their correct synodic periods; as we have seen (pp. 187–88), this was not possible on the original theory. Schiaparelli is able to show that if the fifth sphere, carrying the planet M on its equator (see Fig. 12), has its pole (P_2) fixed to the under side of the fourth sphere (the pole of which is P_1) at a

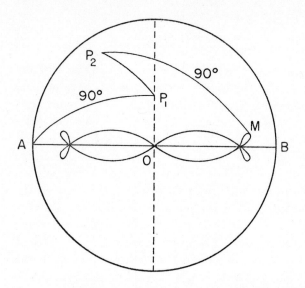

Fig. 12. Possible effect (as conjectured by Schiaparelli) of Callippus' modification of the Eudoxan scheme for Mars, Venus, and Mercury (cf. Heath, Arist. p. 214)

certain angle of inclination (represented by the arc P_1P_2), and rotates at the same speed and in the same direction as the third (which itself rotates at half the speed of the fourth and in the opposite direction and has a period equal to the synodic period of the planet), then a curve is produced which looks like a hippopede, but has two additional small loops at each extremity with two triple points, in addition to the double point (o) at the centre, all of which lie on the ecliptic AB. If P_1P_2 is taken as 45° and the correct synodic period for Mars, 780 days, is assigned to the rotation of the third sphere (i.e. of P_1 round AB), it can be shown[361] that the construction will, in fact, produce a retrograde movement for the planet and a deviation in latitude which approximate to the facts. With similar assumptions for Venus, its greatest elongation becomes $47\frac{2}{3}$° (nearly correct), and since the time taken from one triple point to the other amounts to one fourth of the total and the passage back again to another fourth, while the remaining half of the time is taken up by the very slow motion round the two small loops at each end, the construction goes some way towards explaining the different speeds of the planet in moving from greatest eastern to greatest western elongation and vice versa (see p. 188). The fifth sphere for Mercury may have been used in similar fashion to bring the theory closer to the planet's observed motions.

Thus in certain respects the modifications introduced by Callippus were (or rather, could have been) improvements on the original Eudoxan system, but the fundamental deficiencies remained the same. Since he accepted the same number of spheres as Eudoxus for Saturn and Jupiter, it is evident that he also had no conception of the zodiacal anomalies of these planets, and therefore almost certainly not of the other planets as well. There is no hint of planetary orbits inclined to the ecliptic, and the Callippic system is no more successful than the Eudoxan in accounting for the noticeable differences in brightness of the planets (and hence in their distance from the earth). Understanding of the motion of the moon was advanced to some extent, and of the sun to a greater extent, as is evidenced by Callippus' comparatively accurate values for the astronomical seasons; but since he retained the three spheres of Eudoxus (adding two more of his own), he apparently still believed in the fictitious latitudinal deviation of the sun from the ecliptic which Eudoxus introduced (see above, p. 182). Callippus' concern with solar observations (presumably of solstices and equinoxes) is borne out by his success in correlating more accurately the lunar month with the solar year. He improved on the Metonic intercalation cycle of 19 years by taking four such cycles less one day (i.e. 27,759 days), in the course of which there were 940 lunar months (including 28 intercalary ones), to form one 'Callippic period' of 76 years.[362] This gives a solar year of exactly $365\frac{1}{4}$ days[363] (in fact, the Julian year, which is some 11 minutes too long) and a mean lunar month of 29 days 12 hours 44 minutes 25.5 seconds (only about 22 seconds too long). Ptolemy in the *Almagest* frequently uses Callippic periods in conjunction with Egyptian reckoning for dating astronomical observations,[364] and comparison of these dates shows that the first Callippic period began in 330 BC, probably on 28 June when there was a new moon at the summer solstice.[365]

We come next to the second giant of Greek philosophy, Aristotle (384–322 BC), who was born at Stagira in Thracian Chalcidice, studied under Plato in Athens from 367 until the latter's death in 348, and then, after periods of residence at Assos in Mysia, Mytilene in Lesbos, and Pella in Macedonia (where he was tutor to Alexander the Great), returned to Athens in 335, and founded his own school of philosophy, the Lyceum, also known as the Peripatetic school (from περίπατος, a covered walking space). Aristotle, like Plato, was not primarily an astronomer, and his astronomical opinions have to be gathered from largely incidental references in the extant corpus of his works. Since these include his treatise *On the Heavens* (Περὶ οὐρανοῦ, *De Caelo*[366]), it might be expected that this would be an easy task. However, the *De Caelo* is by no means a work on astronomy proper; rather it is a philosophical discussion of certain, selective, cosmological principles treated in the abstract, to show how the basic tenets of astronomical thought can best be accommodated to the general Aristotelian world-picture in line with the rest of his philosophical approach. In fact, for information about such matters as the order, relative positions, distances, and relative motions of the stars (i.e. the main concerns of scientific astronomy), he himself expressly refers us to the astronomers.[367]

The difficulty lies to a great extent in the nature of the works of Aristotle that have come down to us. As is well known, of the three classes of writings connected with his name, viz. (*a*) early dialogues in the Platonic manner, (*b*) collections of material and memoranda for detailed scientific studies, and (*c*) philosophical and scientific treatises, it is only the last class that is represented by the extant works (except for the *Constitution of the Athenians*, Ἀθηναίων Πολιτεία, which belongs to class (*b*)); of the other two classes only fragments remain. However, the surviving examples of class (*c*) are not finished literary products intended for publication in their present form and each presenting a consistent body of homogeneous doctrine in its particular field (like a dialogue of

Plato's). Too full to be regarded as mere lecture notes, they read more like summaries of lectures and discussions or 'aides-mémoires' designed to recapitulate Aristotle's thinking on particular topics. As such, they were frequently added to at different times, occasionally reworked by him, and loosely assembled under different headings, without much regard to internal consistency of doctrine (we shall notice later some striking inconsistencies in his astronomical opinions). Hence a treatise like the *Physics* or *Metaphysics* or *De Caelo* cannot be regarded as a finished whole, nor as representing Aristotle's thought at a well-defined stage of his philosophical development. Individual books, even individual chapters, were written at different times and incorporated, either by Aristotle himself or by later editors, in contexts where they seem out of place, either because they interrupt the unity of structure or because they are juxtaposed with passages where a different line of thought is being pursued. Thus (to give a few examples pertinent to his astronomical ideas) *Physics* vii seems out of place where it now stands and was probably written earlier than viii which was meant to supersede it.[368] *Metaphysics* Λ is generally considered to have been originally a separate work inserted into the body of this treatise (like Books α, Δ, and K) and to be one of the earliest written parts;[369] but ch. 8, which mentions Callippus' modifications to the Eudoxan theory, and is written in a fuller style compared with the short, pithy sentences of the rest, is thought to be later than the rest of the book. Within this chapter, however, 1074a31–38 (which interrupts the argument and leaves οὗτοι in 1074b3 without an antecedent) is again in the abbreviated style, and was presumably inserted from the earlier part.[370]

Most of the *De Caelo* (and this applies particularly to Books i and ii which are the ones chiefly relevant to a history of astronomy) is generally regarded as early (written, for example, before the development of the theory of the Unmoved Mover in *Phys.* vii and viii), but there is a good deal of evidence to prove that Aristotle added to his original arguments piecemeal from a later and different standpoint, without troubling to reconcile divergent views arising from his own changing outlook.[371] All this makes it difficult to determine not only the details (which he is, anyway, content to leave to the astronomers – see above), but also the basic concepts that underlie his astronomical beliefs at various stages of his thought.[372] In what follows, since space precludes the detailed analysis of the *De Caelo, Physics, Metaphysics, Meteorologica, De Generatione et Corruptione,* and *De Anima* (not to mention parts of the *Parva Naturalia*), which would be necessary for a thorough discussion of Aristotle's astronomy in relation to his philosophical concepts, I confine myself to a general outline, drawing attention to those of his views which diverged from his predecessors and those which were destined to have a decisive influence (for good or ill) on his successors.

To begin with, it is certain that Aristotle accepted the concept of the celestial sphere with the spherical earth resting immovably at the centre and the heavenly bodies moving in circular orbits round it; this, as we have seen, was a comparatively new concept, but from the fourth century BC onwards it was never doubted by any serious writer.[373] In the *De Caelo* Aristotle devotes the longest chapter in the whole work (ii, 13) to discussing the position and shape of the earth, reviewing and criticizing in some detail the theories of previous thinkers,[374] and he follows this up in the next chapter by proving that the earth cannot be in motion and must be a sphere. His arguments are largely *a priori*, based on his doctrine of natural motion (on which, see below), but he supports them by appealing to the facts of observation, and since observation plays for the most part a very small role in this treatise,[375] it is worth noticing what use Aristotle makes of it. He says (*De Caelo* ii, 14, 296a34ff.) that:

all the bodies that are carried by the cyclic revolution [i.e. the planetary bodies including sun and moon] are seen to lag behind and be moved by more than one revolution apart from the first one [i.e. the diurnal revolution of the sphere of the fixed stars – Aristotle is referring to the fact that, in the modified Eudoxan system which he adopted (see below), all the planetary bodies have at least three spheres of motion][376]; hence the earth also, whether it is carried round the centre [as in the

Pythagorean Philolaic system] or rotates at the centre [according to another, unnamed school of thought],[377] must necessarily have two revolutions [in addition to one of the above]. If this occurred, however, there would necessarily be deviations and turnings (παρόδους καὶ τροπάς) of the fixed stars; yet this is not seen to happen, but the same stars rise and set at the same places on the earth.

Arguing against the proponents of an earth with any sort of motion, Aristotle points out that, by analogy with the other planetary bodies, a moving earth would have to move in such a way that, to an observer on it, the stars would seem to move as the planetary bodies do at present for an observer on a stationary earth – i.e. with deviations in latitude (παρόδους) and 'turnings' (τροπάς),[378] i.e. the points where the body reaches its maximum distance from the equator for the sun (namely, the solstices) and from the ecliptic for the moon and planets. However, courses like this are *not* observed for the stars, and therefore the earth cannot be in motion. Another proof adduced on observational grounds is that 'the changes of the figures by which the arrangement of the stars is defined' – by which Aristotle means the regular succession of different constellations over the horizon at different times of the year repeated annually, not changes in the actual shapes of these (which he well knew remained constant, ii, 6, 288b12) – are consistent with an immovable, central earth (297a4–6).

The sphericity of the earth is proved (*a*) by *a priori* arguments based on the assumption of the natural movement downwards of all earthy matter towards a centre (hence the final result must be a sphere, the figure that includes the maximum volume of evenly distributed matter round the centre – 297a8–b23), and (*b*) by the observation that the demarcation-line between the bright and dark portions of the moon's disc during a lunar eclipse is always convex.[379] He adds that since a relatively small change in the position of the observer northwards or southwards makes an appreciable difference in the stars visible above the particular horizon,[380] the earth's sphere cannot be very large in relation to the stars (*cf. Meteor.* i, 3, 339*b*9); so that those who argue for the junction of Africa and India by a single ocean because elephants are found in both, are not so wide of the mark (298*a*9f. – a statement that is said to have influenced Columbus; *cf. Meteor.* ii, 5, 362*b*18). Aristotle then mentions that 'the mathematicians who try to measure the size of the circumference' (perhaps referring either to Archytas – see above, p. 76 – or to Eudoxus – see my *GFH*, p. 24) put it at 400,000 stades; this is equivalent to 46,000 miles, nearly twice the true figure of about 24,900 miles,[381] and is the oldest estimate of the earth's circumference that has come down to us.

Not only the earth, but all the celestial bodies are spherical in shape (ii, 8, 290*a*7). Observation shows that the moon must be a sphere, because, if it were any other shape, its phases would not assume the form they do, nor would the sun appear crescent-shaped during a solar eclipse; and if one heavenly body is spherical, then all the others will be too (ii, 11, 291*b*18f.). Whatever we may think of this last argument, it is at any rate clear that Aristotle knew the correct explanation of the moon's phases and of solar and lunar eclipses (*cf. Met.* H 1044*b*9; *Anal. Post.* ii, 2, 90*a*15). Moreover Nature makes nothing without a purpose, [382] and the sphere is the shape that is least suited to progressive motion (Aristotle does not seem to have studied ball-games . . . *cf.* Randall, *op. cit.*, p. 126), but the best suited for maintaining its own position (because the heavenly bodies do not move by themselves, but are carried round in different spheres – 289*b*32 and see below). Hence it is clear that all the stars (including the planetary bodies) must be spherical masses (ii, 8, 290*b*1ff.; 11, 291*b*11ff.). In contradiction to Plato (*Tim.* 40*a*–*b*), Aristotle argues against any rotatory movement of the stars; the fact that the sun seems to rotate at its rising or setting[383] he explains by the weakness of our vision since we see it from a great distance, and this is also why the stars seem to twinkle while the planets (which are nearer to us) do not (290*a*14–23). Aristotle's reasoning here is unsatisfactory, not only because in his system the sun is nearer the earth than all the planetary bodies except the moon, but also because, although he was aware that the moon always shows the same face to the earth (290a26), he apparently still denies its rotation.[384] The great distances and sizes of the celestial bodies are emphasized in *De Caelo* ii, 9, where he mentions only to reject the Pythagorean theory of the harmony of the spheres (see above, pp. 71–2).

For Aristotle, the universe is finite (*De Caelo* i, 5, 6 and 7), there is no void (*Phys.* iv, 214a16ff.), there is only one world (*De Caelo* i, 8 and 9),[385] and that is ungenerated and indestructible (*De Caelo* i, 10, 11 and 12). Thus he rejects the views of Democritus and Leucippus (void and more than one world), as well as Plato (creation of the world by the Demiurge) and Heraclitus (cyclic birth and destruction of the world); and in *De Caelo* ii, 1, 284a24f., he discards the vortex (δίνη) of Empedocles and Anaxagoras. Aristotle's own views on the nature and composition of the universe are set out in *De Caelo* i, 2 and 3. Accepting the by now standard doctrine of the four elements (earth, air, fire, and water) as the basic constituents for all terrestrial matter, he postulates a fifth element, aether[386] (the famous *quinta essentia* of medieval philosophy), or as he more frequently calls it, the 'first body' (πρῶτον σῶμα), as the basic constituent of the celestial regions. This first body is uncompounded, ungenerated, eternal, unalterable, neither heavy nor light, and the most divine element in the universe. It is not of equal purity throughout the whole of the celestial regions, but is least pure where it borders on the air round the sublunary region (*Meteor.* i, 3, 340b8–11). Its natural motion is motion in a circle which it carries out continuously (as evidenced by the movement of the heavenly bodies), in contrast to the limited rectilinear motions of the other elements, e.g. of fire, the natural motion of which is straight up away from the centre, and earth, straight down towards the centre (cf. *Phys.* v, 230b12f.). Moreover, the place of the first body, the heavens, where the fixed stars, planets, sun, and moon carry out their unceasing revolutions, is sharply differentiated from the sublunary world of continual change which in all its parts exhibits the processes of generation, growth, diminution, and decay (cf. *Meteor.* i, 14). The idea of the fifth element (already present in Plato as the 'brightest and clearest' portion of air – εὐαγέστατον,

Tim. 58*d*), and the distinction between the celestial regions and the sublunary world, may be seen in some respects as a logical outcome of Plato's insistence on the divinity of the heavenly bodies and the perfection of their courses (*cf.* pp. 140–41; 104). Certainly it had an enormous influence on the thinking of the Middle Ages where it became an integral part of the world picture developed on the basis of a monotheistic religion, and provided, for example, a sphere of action for the spirits and angels of medieval religious concepts.

We come now to the only passage where Aristotle goes into any detail (and this is little enough) of how he envisaged the arrangement of the planetary spheres, *Met.* Λ 8. After describing the Eudoxan system and Callippus' modifications (see above), which he evidently accepted, he continues (1073*b*38–1074*a*5):

> It is necessary, if all the spheres put together are going to account for the observed phenomena, that for each of the planetary bodies there should be other counteracting [literally 'unrolling', ἀνελιττούσας] spheres, one fewer in number [than those postulated by Callippus for each set] and restoring to the same function each time as regards position the first sphere of the planetary body situated below; for only thus is it possible for the whole system to produce the revolution of the planets.

The purpose of these counteracting spheres is further explained by Sosigenes in Simplicius (pp. 498*ff.* ed. Heiberg – *cf.* Heath, *Arist.*, pp. 217*f.*, based on Schiaparelli): it is to cancel out the motions of all but the first sphere (representing the daily revolution of the whole heavens common to all the planetary bodies) in each set, so that the next planet down can carry out its motions uninfluenced by the individual movements of the previous one. Thus for the four spheres of Saturn, A,B,C,D, a counteracting sphere D′ is postulated, placed inside D (the sphere nearest the earth and carrying the planet on its equator) and rotating round the same poles and with the same speed as D but in the opposite direction; so that the motions of D and D′ effectually cancel each other out, and any point on D will appear to move only according to the motion of C. Inside D′ a second counteracting sphere C′ is

placed, which performs the same function for c as d′ does for d;
and inside c′ is a third counteracting sphere b′ which similarly
cancels out the motion of b. The net result is that the only motion
left is that of the outermost sphere of the set, representing the
diurnal rotation, so that the spheres of Jupiter (the next planet
down) can now carry out their own revolutions as if those of
Saturn did not exist. In the same manner, Jupiter's counteracting
spheres clear the way for those of Mars, and so on (the number of
counteracting spheres in each case being one less than the original
number of spheres in each set) down to the moon which, being
the last of the planetary bodies (i.e. nearest the earth), needs,
according to Aristotle, no counteracting spheres.[387] He goes on
to state that the total number of spheres (including counteracting
ones) in the system is 55.

The following table gives a comparison of the numbers of
spheres envisaged by Eudoxus, Callippus and Aristotle, always
excluding the sphere of the fixed stars itself with its single motion
at the extremity of the universe:

	Eudoxus	*Callippus*	*Aristotle*
Saturn	4	4	4 + 3 counteracting
Jupiter	4	4	4 + 3 ,,
Mars	4	5	5 + 4 ,,
Venus	4	5	5 + 4 ,,
Mercury	4	5	5 + 4 ,,
Sun	3	5	5 + 4 ,,
Moon	3	5	5
	26	33	55

Aristotle then adds: 'But if one does not add to the sun and moon
the movements I have mentioned, the total number of spheres
will be 47.'[388] This remark has troubled commentators both
ancient and modern, because it seems to convict Aristotle of a
mistake in elementary arithmetic. If by ἃς εἴπομεν he means only
the 4 counteracting spheres he himself added for the sun, the
total should be 51. Or is he including another 4 for the moon,

having forgotten that he has just said that it does not need any? If he means the spheres added by Callippus (2 each for sun and moon) plus 2 counteracting ones for the sun (this would seem the most logical interpretation of his words), the total ought to come to 49. Or is he discounting Callippus' extra solar and lunar spheres *and* omitting his own 4 counteracting spheres for the sun – which would in fact give 47 – forgetting that even with only 3 spheres for the sun, 2 counteracting ones are still necessary in accordance with his own views? If we are to absolve Aristotle from making a slip, the only alternative is to suppose that ἑπτά is a scribal error for ἐννέα, as was in fact suggested by Sosigenes (*ap.* Simpl., p. 503, 35). It is at any rate clear that Aristotle was in some doubt about the total number of spheres to be postulated – *cf.* Simpl., p. 505, 23*ff.*

Dreyer (p. 113) and Heath (*Arist.*, p. 218–19), followed by Ross (*Arist. Met.*, p. 392), state that Aristotle might have dispensed with 6 of his 55 spheres, because the innermost counteracting sphere (B′ in the above example) moves in the same way as the sphere of the fixed stars, which is also the motion of the first sphere in the next set down; so that these 2 spheres might have been replaced by 1, and this could have happened in the case of each of the six planetary bodies after Saturn. This criticism, however, is mistaken. B′ in order to cancel out the motion of B, must have the same axis of rotation and move with the same speed (but in the opposite direction) as B. But the poles of B are *not* the poles of the sphere of the fixed stars (see above, p. 178), whereas it is an essential part of the system that the first sphere of each set must represent the latter exactly; hence the 2 spheres *cannot* be replaced by 1. This is why Aristotle himself emphasizes that the purpose of his counteracting spheres is to 'restore to the same function *as regards position*' the first sphere of the following planetary set.[389] In fact Dreyer and Heath make the same mistake as some of the ancient commentators, who thought that the last of the counteracting spheres was the first of the activating spheres of the next planet, and who therefore had to count the same sphere twice in order to retain Aristotle's total – an error corrected by Sosigenes with Simplicius' approval (p. 502,20). Thus a total of 55 spheres *is* necessary for the complete Aristotelian system[390] – at least in theory. Obvious difficulties arise if we enquire too closely into the actual physical connection of the spheres. For example, if the heavens really operated in this manner, with the counteracting spheres effecting their respective cancellations of planetary motions, how did astronomers ever manage to make the observations that lay behind the original Eudoxan scheme with its planetary loops and retrogradations? Perhaps this is why Aristotle spends so little time on the ἀνελίττουσαι σφαῖραι, which he dismisses in a dozen or so lines and never refers to elsewhere in his works; it is really his commentators who have elaborated the concept, which Aristotle may merely have considered an interesting speculation, but one that would not stand up to close scrutiny.

This complicated system is to a great extent forced on Aristotle by his mechanistic view of the structure of the universe and his opinions on the nature of movement in general. He apparently regards the whole universe as made up of spherical shells in physical contact with each other. In *De Caelo* ii, 4, starting from the acknowledged fact of the sphericity of the heavens (the sphere being the most perfect solid figure and therefore most appropriate for the first body with its natural, circular motion), he shows that the realms of the other elements (right down to the central, spherical earth), which are in continuous contact with each other and the heavens,[391] are also spherical, proving this in the case of water (the element immediately round the earth, with air next, and fire round air) by its property of 'always running together into the hollower place – and "hollower" means nearer the centre' (287a32–b7). The fifth element is corporeal (i, 8, 277b13–17) and occupies the region down to the moon, becoming less pure where it borders on the sublunary region (*Meteor.* i, 3, 340b4ff.). The heavenly bodies are made out of it and so are the circles (or rather spheres) in which they move (*De Caelo* ii, 7, 289a13–16); for the stars (including the planetary bodies) do not move of themselves, but are 'fixed into' (ἐνδεδεμένα, ii, 8, 289b33) their particular spheres (regarded as also corporeal, ii, 7, 289a28–32)[392] and carried round with them – just how, Aristotle wisely makes no attempt to explain (*cf.* Moraux, p. xlvi). Thus, in a planetary set, the last sphere which carries the planet is described as 'fixed into and carried by many spheres and each sphere is corporeal.'[393]

In *De Caelo* ii, 3 Aristotle explains why it is necessary to postulate other circular motions for the planetary bodies besides the divine, eternal motion of the first heaven (οὐρανός in the first of the senses just mentioned). The reason is that there must be a centre of circular movement and this centre is the stationary earth (proved in ii, 14); but if earth exists, so must fire (for these are contraries, and the existence of one of a pair of contraries entails the existence of the other) and the intermediate elements, water and air (*cf. De Gen. et Corr.* ii, 3 and 4). Now if there were only the one circular movement of the whole, the relation of the elements to each other would remain unchanged (each remaining in its particular place – see above); but in the sublunary region there is constant interchange between them and continual 'coming-to-be' (γένεσις) and decay (φθορά). Therefore there must be one or more circular motions besides that of the first heaven, in order to make 'coming-to-be' possible. In fact, it is the motion of the sun in the ecliptic (ὁ λοξὸς κύκλος), producing warmth and light and the alternation of the seasons, which is the cause of all terrestrial life.[394]

The symmetry of this conception (the eternal, unchanging motion of the first heaven corresponding to the eternal, fixed earth, and the continuous but changing motion of the sun corresponding to the continual changes in the sublunary region – *cf. Met.* Λ 6, 1072a 9–17 with Ross' note *ad loc.*) evidently appealed to Aristotle's tidy mind. He would have liked to have found further symmetry in the arrangement of the planetary motions, which might have been expected to increase in number in proportion to the distance from the single motion of the first heaven (*De Caelo* ii, 12, 291b28ff.). But, in fact, the earth is stationary, and 'sun and moon move with fewer motions than some of the planets',[395] although the latter are nearer the first heaven. This remark has caused difficulty, because, as a glance at the table above will show, in none of the three systems have the sun and moon fewer spheres than *some* of the planets, although in the original Eudoxan scheme they have fewer spheres than *any* of the planets. Probably Aristotle is not here thinking of the theoretical spheres, but only of the apparent movements in the sky (κινήσεις not σφαῖραι being mentioned); sun and moon do not exhibit loops and intermittent retrograde motions along the ecliptic, nor in the original Eudoxan system apparently do Mars and Venus (see p. 187), and it is possible to suppose that at this time Aristotle had not yet adopted Callippus' modifications or the theory of counteracting spheres. Hence it would be correct to say that sun and moon have fewer motions than *some* of the planets (e.g. Saturn, Jupiter and Mercury).

Aristotle's views on motion in general (used in the broadest sense to include change of place, quality and quantity – cf. *Phys.* vii, 243a35–7) are set out in *Physics* v–viii; for astronomy we are chiefly concerned with the first of these categories, spatial motion (locomotion), which he regards as primary (viii, 7). Some things are always at rest (i.e. the earth at the centre of the universe), others are always in motion (i.e. the celestial bodies), while others are sometimes at rest and sometimes in motion (i.e. the things of the sublunary region) – cf. *Phys.* viii, 3 and 4, especially 253a28–30. Circular movement is eternal, continual, and without beginning or end (viii, 8). Everything that is in movement is moved by something, the mover must be in contact with the moved,[396] and in any series of movements there must be a first mover that is itself unmoved (vii, 1; elaborated in viii, 4). Aristotle is insistent that this applies to *all* movers and *everything* moved, which therefore includes the heavenly bodies (cf. viii, 6, 259b29–31) – hence the mechanics of the system of planetary spheres (see above). There must, then, be an unmoved first mover, which is unaffected by anything else and unmixed, and Anaxagoras was right to describe 'mind' (νοῦς) in these terms since he made this the first principle (ἀρχή) of movement (viii, 5, esp. 256b23–7). The unmoved first mover (or movers, since it makes no difference to the argument here whether there is one or more than one[397]) is eternal because the movement it initiates is eternal (viii, 6), and it is the ultimate source of all movement in the universe in which motion (of some kind) is continuous, imperishable, and has always existed (viii, 1 and 2). Moreover, the first mover is without parts or magnitude; it must be situated either in the middle or at the circumference (because these are the 'generative principles' of the sphere[398]), and since things move quickest that are nearest to the mover, and the circular motion of the circumference (i.e. the sphere of the fixed stars) is the quickest, then the first mover must be at the circumference (viii, 10, 267b6–9).[399]

So far we have seen the case for an unmoved first mover (henceforth I use the term 'prime mover' both singular and plural) argued on general philosophical grounds, though with hints that its prime function is to explain the revolution of the celestial sphere. In *Met.* Λ 6–8 Aristotle elaborates the concept and develops its astronomical application. A prime mover is one of the three universal types of entity or essential substance, οὐσία (the other two are the perishable things of the sublunary region and the imperishable but observable celestial bodies; cf. 1069a30f.); it is pure actuality (ἐνέργεια), everlasting and without matter,[400] and there is no need to seek any other principles of movement, such as the Platonic Ideas or the concept of soul, than these prime movers (ch. 6). The first heaven (πρῶτος οὐρανός) exhibits everlasting circular movement; there must therefore be an entity (οὐσία) that moves it, and this entity must be itself unmoved and everlasting. It operates by acting as the goal for desire and thought (κινεῖ δὲ ὧδε τὸ ὀρεκτὸν καὶ τὸ νοητόν, 1072a26; cf. *De Mot. Anim.* 6,700b36). Being itself utterly and unchangeably good, the first and best entity in the universe, it causes the motion of the primary form of movement, that of revolution in a circle, which strives after it as a lover strives after the loved object;[401] and on such a principle (ἀρχή – a prime mover is an ἀρχή, 1072b10–11) the whole physical universe depends. The prime mover's activity is the highest form of joy (ἡδονή) which is pure contemplation with itself as object, such as we mortals very rarely attain to, but which is the natural state for the divine; the prime mover is a living entity (ζωή), the best possible and eternal, and since these are attributes of the god, it is itself divine;[402] it is immaterial (κεχωρισμένη τῶν αἰσθητῶν) and has no parts and no magnitude (ch. 7).

Aristotle then poses the question whether one or more such

prime movers must be postulated, and if the latter, how many. After pointing out that the doctrine of Ideas is of no help here, he states in the most explicit manner that there must be a plurality of prime movers. A single prime mover is responsible for the single, everlasting, circular movement of the whole heavens, but, since we observe that there are other eternal, circular motions of the planetary bodies (each of which is itself an eternal entity – ἀίδιος οὐσία – in everlasting motion), each of these motions also must have its own prime mover. All these prime movers will have the same characteristics as set out above, and they will be counted as first, second, third, etc., according to the arrangement of the planetary revolutions (8, 1073a26 – b3). Their number must be ascertained from astronomy (ἀστρολογία – preferred by Aristotle to its synonym, ἀστρονομία, and used with none of the connotations of the later 'astrology'), which is 'the closest of the mathematical sciences to philosophy' because it deals with entities (οὐσίαι) that are both perceptible and eternal, whereas sciences like arithmetic and geometry are not concerned with entities at all (1073b8).

There follows the account of the Eudoxan system, Callippus' modifications, and Aristotle's own version with the counteracting spheres (see above), but first Aristotle makes it clear that he goes into such astronomical details reluctantly and only for the sake of having some figures to work on, and he is aware of the provisional nature of these (1073b10–16). After concluding that the total number of spheres is 55 or 47 (see p. 201–02), he says that there must then be the same number of prime movers (described as usual as οὐσίας καὶ τὰς ἀρχὰς τὰς ἀκινήτους, 1074a15; the medieval schoolmen called them 'intelligences'), for every circular revolution in the heavens contributes to the revolution of a star and every perfect entity must be considered an end (τέλος) or final cause of circular movement (by arousing desire or thought in the moved, as explained above). Every mover and every revolution exists for the sake of the moved, and no revolution exists for the sake of another revolution[403] otherwise there would be an infinite regress (1074a25–31). There is only one οὐρανός, for if there were several there would have to be one ἀρχή for each and these would have to be accountable by number (ἀριθμός); but things that can be counted by number must possess matter (ὕλη), whereas a prime mover has no matter and is sheer reality (ἐντελέχεια).[404] Aristotle then again (as at the end of *De Caelo* i, 9) draws attention to the fact that popular belief has from time immemorial invested the heavens with the idea of divinity (the anthropomorphic gods of mythology being added later for practical reasons), and this amounts to a realization, the truth of which he affirms, that the prime movers are gods.[405]

Before attempting to evaluate Aristotle's astronomical views, we should notice what he has to say in the *Meteorologica* about phenomena such as shooting stars, comets, and the Milky Way; since, on his own showing (*Meteor.* i, 1), this treatise is to be placed after the *Physics*, *De Caelo*, and *De Generatione et Corruptione*,[406] it might be expected to confirm or modify opinions he expresses in the earlier works. Observation plays a much greater part in the *Meteorologica* than in the *De Caelo*, and there is far less discussion of basic philosophical principles; but the observations are not the kind on which mathematical astronomy is based (*cf.* note 375) and, to a modern reader, have a distinctly 'amateurish' air. In chapters 6 and 7 Aristotle cites records of four comets: in 468/7 BC accompanying the fall of a meteorite at Aegospotami (344*b*32–4);[407] in 427/6 at the time of the winter solstice during the archonship of Euclees at Athens (343*b*5–7); in 373/2 'about the time of the earthquake in Achaia and the tidal wave' in the archonship of Asteius;[408] and in 341/0 in the archonship of Nicomachus (345*a*2–5). Arguing against theories that comets are planets or the conjunction of two planets (the view of Anaxagoras and Democritus, 342*b*27), Aristotle asserts that not only the Egyptians but he himself has observed that some of the fixed stars have tails; for example, 'one of the stars in the thigh of the Dog had a tail, though a faint one – for when one gazed at it intently the light became dim, but when one glanced aside slightly it came into better view' (343*b*12–14).[409] In the same context, he states that he has observed the occultation of one of the stars in Gemini by the planet Jupiter (343*b*30–3). At *Meteor.* ii, 5, 362*b*10 there is a reference to the Crown's (i.e. Corona

Borealis) being directly overhead when on the meridian, and according to Baehr's tables (p. 56) the declination of α Coronae in 350 BC was + 36.6°, so that it would have been almost directly overhead at Athens (lat. 38°N.); but the passage is irrelevant where it stands and is regarded as an interpolation (cf. Lee, p. 180 note a).

To account for phenomena like shooting stars, the aurora,[410] and comets, of which Aristotle distinguishes several types (i, 4 and 7), he explains that the earth when heated by the sun gives off two kinds of exhalation (ἀνθυμίασις), one hot and dry that rises to the top stratum of the sublunary region immediately beneath the motion of the heavenly bodies, and the other vaporous and moist (ἀτμίς) that directly encircles the earth and in which the formation of clouds takes place; the hot and dry stratum is highly inflammable (and therefore is called 'fire', although it is not really fire – 341b15) and under certain conditions it actually bursts into flame, either by contact (presumably frictional) with the celestial motion, or by the ignition of a suitable local condensation of the lower atmosphere expelled upwards, so as to produce the above phenomena (i, 3 and 4, summed up in 7 init.; cf. ii, 4, 359b28; 360a21; ii, 9, 369a13). When a comet is produced by the latter process it is a local phenomenon (i, 7, 344a34), but when it is produced by the action of a planet or a star, it is carried round with the star's (daily) revolution (344b9).[411] In i, 9 (cf. ii, 2, 354b24f.) Aristotle says that the moist vapour round the earth is to be regarded as a circular river that ebbs and flows with the course of the sun, and suggests that this is what the ancients meant by ὠκεανός, the 'ocean river'.[412] As regards the Milky Way (τὸ γάλα), after first showing the inadequacies of his predecessors' views (his normal method of procedure and one for which a historian of science must be duly grateful), he suggests an explanation similar to that of comets, i.e. the ignition of the uppermost stratum of the air by the celestial motion; for, just as a single effect of this kind produces a tail on a single star, so when a whole circle is affected (and it is several times referred to as a 'greatest circle', μέγιστος κύκλος – 345a33; 346a17; 346b7) the tail manifests itself as the phenomenon of the Milky Way (i, 8).[413]

He criticizes previous opinions on the causes of the 'turnings' (τροπαί) of the sun (i.e. the solstices) and moon (ii, 1 and 2); there were two views, (a) that they are caused by evaporated air and wind (353b7; 355a22–5), and (b) that the sun is fed by moisture drawn up from the earth and when it is over certain regions this nourishment fails, so that the sun has to turn back (354b34–355a5). The fact that Aristotle has to argue seriously against these fantastic views (held by several of the Pre-Socratics – see Chapter III) may serve to remind us how near he is to the very beginnings of scientific astronomy.

The fifth chapter of Book ii shows him fully conversant with the type of information contained in the 'parapegmata'.[414] For example, in 361b36, he says 'The Etesian winds blow after the (summer) solstice and the rising of the Dog Star' (referring to a time at the end of July). He also indicates some knowledge of the zones of the terrestrial sphere as delineated by the tropics and what he calls the 'wholly visible circle'.[415] However, his subsequent remark (362b6), that the regions beyond the tropics are uninhabitable because the earth ceases to be habitable where the (midday) shadow first begins to fall towards the south (i.e. from the tropic of Cancer, probably placed by him at the equivalent of 24°N. – see above, pp. 157–58 – southwards), reveals his deficient knowledge of more southerly latitudes. By contrast, the statement that the regions 'under the Bear' (ὑπὸ τὴν ἀρκτόν, i.e. where the Great Bear is directly overhead on the meridian – say, north of about 64°) are uninhabitable because of the cold is more or less correct. In one respect in the *Meteorologica*, Aristotle has changed his opinion from that expressed in the *De Caelo*. In the latter work (ii, 2, 285b15) he is forced to conclude that the visible pole is the lower (i.e. southern) one;[416] but in *Meteor.* ii, 5, 362a33 he abandons this artificial notion and agrees that the pole in our hemisphere is the upper, northern one.

When one comes to the *De Caelo* and the *Meteorologica* fresh from reading the *Organon* or the biological treatises, it is difficult to avoid a sense of disappointment, a feeling that instead of experiencing the play of a powerful intellect on problems eminently suited to it, one is witnessing the fumbling efforts of a thinker who is not quite sure how to deal with his admittedly intractable subject-matter. The incisive logic that analyses so expertly the processes of reasoning is missing, as is also the meticulous observation and careful classification that is a feature of Aristotle's other works. This cannot be explained entirely by the fact that Aristotle was not a professional astronomer, or by any lack of interest on his part in astronomy. Neither was he a professional politician or literary critic, and yet the *Politics* and *Poetics* do not give this impression of uncertainty and of only a halting progress towards an ill-defined goal. His interest in astronomy (which he shared with most of the Greek philosophers) was lifelong, as is attested by the fragments of the early dialogue *De Philosophia* (particularly fragments 10–12, 14, and 18), and, like Plato's, seems to have increased in his later years.[417] Yet this interest is less in astronomy as a science than as an adjunct to philosophy (*cf. Met.* Λ 8, 1073b4–5), an attitude which is perhaps not surprising in view of his well-known dislike of mixing too much mathematics with philosophy.[418] In some respects it may be said that astronomy (but not mathematical astronomy, which he left to the professionals – see also above, p. 194) plays the same role in Aristotle's system as mathematics does for Plato;[419] but it is an astronomy that must conform to his basic philosophical principles and be consistent with his teleological and mechanistic universe. Contemporary astronomical views, as Plato had recommended (see pp. 107–08), took more account of the mathematical relationships governing the movements of the celestial bodies than their physical characteristics as parts of a unified structure. Aristotle found it difficult to reconcile them with his own ideas on the nature of the universe, and this is responsible for the inconclusiveness and inconsistencies of his astronomical opinions. For it must be admitted that in this field we see Aristotle at his least convincing.

Perhaps the greatest inconsistency relates to his concepts of the fifth element and the prime movers. To put it at its simplest: if the natural motion of the fifth element (the constituent material of all the heavenly bodies and their spheres) is movement in a circle, which it carries out continuously, of its own accord, and unchangeably,[420] why is it necessary to postulate any prime mover to initiate this movement? The favourite explanation (e.g. Guthrie, Loeb *De Caelo*, pp. xix*ff.*) is to assume that the concept of prime movers is a late development in Aristotle's astronomical thought, which had not occurred to him when he put forward the idea of the fifth element. The fact that there are several passages in the *De Caelo* which unmistakably *do* refer to a prime mover[421] is then explained by assuming that these are later insertions added without regard to their consistency with the earlier doctrine (*cf.* pp. 194–95). A persuasive case can be made for such a view with its tidy assumption of recognizable phases of development in Aristotle's astronomical ideas; but there are also strong arguments against it and some convincing evidence that the prime mover theory was always part of his thought, though not developed in detail until his later years (see especially Cherniss, *Arist. Crit. Pl. and Acad.*, App. X).

No attempt can be made here to discuss the relevant passages in detail,[422] but it seems worth mentioning two points to which insufficient attention has been paid. The first is that it is not only the isolated passages referred to above (288*a*27f.; 288*b*22f.) which imply recognition of the role of a prime mover, but the whole of ii, 6. This is the chapter in which Aristotle proves that the motion of the first heaven is uniform and regular; he gives a series of arguments designed to show that irregular (ἀνώμαλος) motion necessarily entails acceleration (ἐπίτασις) and deceleration (ἄνεσις), that these effects are not observed to occur (either for the whole or its parts),[423] and that therefore the revolution of the heavens is not irregular – which, in fact, has already been assumed in ii, 4.[424] These arguments are carefully linked together in style as well as subject-matter (e.g. a keyword at the end of one argument will be repeated at the beginning of the next[425]) and it is hard to believe, when one reads them consecutively, that two of them are later insertions. Besides this, acceleration and deceleration imply a 'climax' (ἀκμή) or time of maximum velocity either at the beginning or at the middle or at the end of the motion, but since revolution in a circle has by its nature no such absolute points, it cannot exhibit any climax (288*a*18f.); this argument could hardly be used without the supposition of a mover to produce the acceleration and deceleration, and in fact two of the examples Aristotle gives (namely, things moving 'contrary to nature', παρὰ φύσιν, and missiles) presuppose a mover (cf. Phys. viii, 4).

The second point to be made is that there is not, in reality, such a deep-seated contradiction between the postulate of the fifth element and the concept of an unmoved mover as is commonly believed. The continuous circular movement of aether (which is its nature) does not really fit in with the rest of the Aristotelian doctrine of natural motion,[426] but is very much sui generis; it has, for one thing, no goal of motion in the sense that 'up' characterizes the natural motion of fire and 'down' that of earth. It would not therefore be surprising if a mover operated on this element in a different way from that which we are accustomed to connect with the other elements; and in fact we are told that its mode of operation is through desire and thinking (see above, p. 206). Now the contradiction that all the commentators find so reprehensible is based on the supposition that the initiation of movement is the prime function of the mover. But suppose that this was not so in the case of the peculiar fifth element—that Aristotle was thinking mainly of accounting for the facts that the motion takes place round certain poles and not others, and at a certain speed and no other, and in a certain direction and not in another[427] —would not this afford a valid reason for the postulate of a prime mover, even though the element on which it worked was already endowed with continuous circular motion? Perhaps the nearest hint we find in the De Caelo that this was at the back of Aristotle's mind is in ii, 5, where he discusses why the heavens should rotate in one direction rather than the other. He has no real explanation to offer, except that it happens that the way it does rotate is the best possible, but in one place he says, 'Necessarily, either this itself [the fact of rotation in a particular direction] constitutes a principle or there is a principle of it.'[428]

The interpretation suggested above also accounts for the necessity of postulating a plurality of prime movers. Aristotle knew perfectly well that the single revolution of the first heaven, although carrying with it all the heavenly bodies and itself sufficient to explain the movement of the fixed stars, could not be regarded as producing the *individual* movements of the planetary bodies, because the spheres that made up the motions of the latter each had different axes of revolution and revolved at different speeds and in different directions; hence each sphere needed a separate prime mover. Ross suggests (*Arist. Met.*, p. cxxxvi) that these prime movers (the 'intelligences' – see above, p. 207) are themselves moved by *the* prime mover (i.e. that of the first heaven) and are actuated by desire of it. He then points out (p. cxl) that, since they are moved by desire of something that they themselves are not, there must be an element of potentiality and hence something quasi-material in them, which is not in *the* prime mover (God); and he complains that the 'intelligences' do not really fit into Aristotle's basically monistic system. However, this criticism is misdirected. Aristotle nowhere says that the 'intelligences' are actuated by the prime mover of the outer heaven.[429] What he *does* say is that all the prime movers have exactly the same nature (*Met.* Λ 8, 1073a32–b3; cf. above, p. 206–07), each being unmoved, everlasting, unchangeable, immaterial, and divine (cf. p. 207). This effectively rules out Ross' interpretation, which is founded on the mistaken assumption that Aristotle's system is essentially monistic and monotheistic.[430]

There are other, more real, inconsistencies in the *De Caelo*. Despite his evident rejection of Plato's World-Soul as the ultimate cause of all movement in the universe (ii, 1, 284a27f. – on this, see Guthrie, pp. xxif.), Aristotle does not seem able to make up his mind whether the celestial bodies have souls or not. In ii, 2, 285a29 he describes the heaven as 'possessed of soul' (ἔμψυχος), and in ii, 12, 292a18f. (cf. 292b1–2) he criticizes the idea of its being 'soulless' (ἄψυχος); but in ii, 9, 291a23 and in the passage just cited (284a27f.) he apparently denies soul to the celestial bodies. Perhaps the best way of reconciling these statements is to suppose that the soul of the stars is so perfectly adapted to their bodily material that there is no conflict between the two (as there is in the case of mortal beings), so that soul and body form one entity to be acted on by a prime mover.[431] Again, there seems to be a clear contradiction between ii, 7 and ii, 9 concerning the manner of movement of the stars and planetary bodies. In the former chapter, Aristotle says that their heat (in the case of the sun) and light are engendered by friction with the air that lies beneath their revolving spheres (289*a*19–21),[432] whereas in ii, 9 he asserts that the stars are one with their medium (*cf.* pp. 203–04), that they do not move in air or fire, and so no noise is produced by their motion (291*a*16*f.*).[433]

Finally, there seems to be some confusion in his ideas about the relative speeds of the heavenly bodies. In *De Caelo* ii, 10 he states that the motion of the outermost heaven is single and is the quickest of all, while the motions of the planetary bodies are slower and more numerous, 'because each one is carried against the (revolution of the) heaven according to its own particular circle'.[434] It is therefore reasonable to suppose, he continues, that the body nearest the single revolution should carry out its own motion in the longest time and the body furthest away from it in the shortest time, since the former is most affected by the revolution of the fixed stars and the latter is least affected,[435] while the bodies in between take times proportional to their distances 'as the mathematicians demonstrate' (291*b*3–10). The implication, then, is that the moon's motion, as the body furthest away from the sphere of the fixed stars and therefore least affected by its contrary revolution, is the quickest. Yet in *Meteor.* i, 3, 341*a*18*ff.*, explaining that the sun's motion is sufficiently rapid and near to the earth to produce heat (this is on the friction theory), he describes the moon's motion as 'close (to the earth) but slow' (κάτω μὲν βραδεῖα δέ), so that it does not produce any heat. It is possible that there is some connection with the views of Democritus, who (according to Lucretius) thought that the moon had the slowest motion (being least affected by the movement of the fixed stars) and was therefore overtaken by the zodiac more quickly than any of the other planets, but to our eyes seemed to move the fastest (see above, pp. 83*f.*); but in any event Aristotle's opinions seem to be muddled on this point.

We see, then, from this outline of his astronomical thought that there is much that he left obscure and many points on which he never arrived at a satisfactory conclusion. This remains true however much allowance we may try to make for the facts that the *De Caelo* is not concerned with scientific astronomy, and that this treatise and the *Physics* and the *Metaphysics* contain opinions representing different stages of his thought.[436] In several places Aristotle himself emphasizes the difficulty of the subject and the provisional nature of the explanations offered, and warns us not to expect final answers to the problems;[437] and we learn also from Simplicius of Aristotle's doubts as to the validity of the whole theory of planetary movements including his own concept of counteracting spheres (Simpl., *in De Caelo*, p. 505,23*ff.*). It would therefore be a mistake to emphasize the shortcomings and inconsistencies and disregard the very real advances in understanding he displays. His acceptance of the concepts of the celestial sphere, of the spherical earth, and of the composite motions of the planetary bodies, his realization that the diurnal rotation of the heavens by itself is not a sufficient explanation of all astronomical phenomena, and even his willingness (despite his own preference for a non-mathematical 'first philosophy') to defer to the authority of 'the mathematicians' (a general term that includes astronomers as well) on astronomical details – all these are signs of a much more sophisticated approach to the problems compared with that of his predecessors.

Despite his rejection of the Platonic Forms and of Plato's mathematical cosmogony in the *Timaeus*, it is obvious that Aristotle's astronomical opinions are greatly influenced by those of his master. In *De Caelo* ii, 2, 285*b*28 he speaks of 'the second revolution, that of the planets', and this expression clearly recalls the language of the *Timaeus* with its two main circles of the Same and the Different. He took over and expanded the Platonic doctrine of the divinity of the heavenly bodies by inventing for them a supra-mundane fifth element with extraordinary properties; and he developed the equally Platonic doctrine of universal principles into the concept of divine, unmoved movers of the celestial spheres.

He has often been described as a retarding influence on the development of astronomy, because his insistence on the special nature of the celestial region with its own peculiar element, in sharp contrast to the sublunary world with its four terrestrial elements, is supposed to have discouraged the application of the normal laws of science to the phenomena of the heavens; while his doctrine of 'natural motion' did little to advance the understanding of mechanics and dynamics. The latter criticism is somewhat better founded than the former, and yet is still only a part of the truth. It was the lack of the idea of the *controlled experiment* that inhibited the advance of Greek physics in general.[438] Compared with this handicap, Aristotle's doctrine may well be defended, on the ground that even an inadequate theory is better than no theory at all. The former criticism will not stand up to close examination. Aristotle himself was well aware that planetary movements must be investigated on a mathematical basis (even though he prefers to leave this to others), as his acceptance of the Eudoxan system amply demonstrates. It is true that his own modification of this by the notion of counter-acting spheres was not very helpful (and did not really convince himself – see above); but there is no evidence that later Greek thinkers were deterred from applying the laws of mathematics to astronomical phenomena because of Aristotle's ontological differentiation between the celestial and sublunary regions. On the contrary, all the evidence points to an increasingly mathe-matical bias in Greek astronomy from the time that Plato first emphasized the necessity for this, culminating in the highly mathematical system of epicycles and eccentrics that we find in Hipparchus and Ptolemy.[439]

The supposition that the concept of the fifth element inhibited astronomical research is certainly false. It could only be true for a science that was mainly concerned with the investigation of the actual physical properties of the stars (which is increasingly the case with modern astronomy), but this is well beyond the scope of naked-eye astronomy. Aristotle himself can hardly be blamed for the fact that some of the main features of his universe (e.g. the unmoved movers, which St Thomas Aquinas interpreted as angels,[440] and the notion of concentric spherical shells) lasted nearly 2,000 years, being incorporated in different forms in the religious beliefs of Mohammedans, Jews, and Christians as official dogma which it was dangerous to question. Moreover, it was not what one might call 'pure' Aristotelian doctrine that appealed to later thinkers and was to prove such a potent influence on medieval thought, but a composite mixture made up of elements from Plato (especially from the *Timaeus*), Aristotle himself (in the *De Caelo* and *Metaphysics*), and the Neo-Platonists (especially Proclus and Plotinus). This blend of doctrines was particularly congenial to the Arabic philosophers such as Avicenna (tenth and eleventh century) and Averroes (twelfth century), who passed it on to the Latin West.[441] If an explanation is needed for the relative stagnation of science during the Middle Ages, it is to be found in this curious amalgam of ideas buttressed by the authoritarianism of religious dogma.

This seems an appropriate point at which to bring this first volume to a close. In many ways Aristotle represents the end of the first stage of Greek scientific astronomy. Theory had assimilated the concept of the celestial sphere, and the mathematics of the circle was being applied with some measure of success to astronomical problems. Observational technique (to judge from Hipparchus' criticisms of Eudoxus' data) was still probably[442] in its infancy and there are no records of any systematic observation for specifically astronomical purposes (as we shall see, the earliest recorded Greek observations are from the third century BC). However the periods of the sun, moon, and planets were known with some accuracy, perhaps from Babylonian sources (see above).

We learn from Simplicius that a modified system of concentric spheres to explain planetary movements held its ground for some time after Aristotle and was accepted by Autolycus of Pitane,[443] an older contemporary of Euclid, *c.* 300 BC; but its deficiencies were recognized at an early stage, and particularly its failure to account for the changes in brightness of the planets, which suggested that their distances from the earth varied considerably (Simpl., p. 504, 25*f.*) – according to Simplicius, Aristotle himself was aware of this (see above, p. 202). The future of mathematical astronomy, however, lay with the development of the mathematical devices of the epicycle and eccentric and the application of these to planetary phenomena; the germ of this idea may probably be seen in the hypothesis put forward by Heraclides Ponticus that Mercury and Venus revolved round the sun as centre (*cf.* Heath, *Arist.*, ch. 18). Now Heraclides was also one of Plato's pupils and a contemporary of Aristotle, but, since his theory heralds the final, definitive stage of Greek astronomy, discussion of it properly belongs to the next volume.

NOTES

ABBREVIATIONS

Abh. z. Gesch. d. Math.	*Abhandlungen zur Geschichte der Mathematik*
AJPh	*American Journal of Philology*
Amer. Journ. Sem. Lang. and Lit.	*American Journal of Semitic Languages and Literatures*
Anth. Lyr. Gr.	E. Diehl ed., *Anthologia lyrica graeca*, 1954
Arch. f. Orientf.	*Archiv für Orientforschung*
Bull. Inst. of Class. Stud.	*Bulletin of the Institute of Classical Studies of the University of London*
CAH	*Cambridge Ancient History*
CQ	*Classical Quarterly*
CR	*Classical Review*

DK	H. Diels, *Die Fragmente der Vorsokratiker*, 6th ed., revised by W. Kranz, 1951–52
Dox. Gr.	H. Diels, *Doxographi Graeci*, 1879
EGP	J. Burnet, *Early Greek Philosophy*, 4th ed., 1930
GFH	D. R. Dicks, *Geographical Fragments of Hipparchus*, 1960
Guthrie, *Hist. of Greek Philos.*	W. K. C. Guthrie, *History of Greek Philosophy*, 1962–
Harv. Theol. Rev.	*Harvard Theological Review*
Heath, *Arist.*	T. L. Heath, *Aristarchus of Samos*, 1913
Heath, *Greek Astron.*	T. L. Heath, *Greek Astronomy*, 1932
Heath, *Hist. of Greek Maths.*	T. L. Heath, *History of Greek Mathematics*, 2 vols., 1921
J. Hist. Id.	*Journal of the History of Ideas*
JHS	*Journal of Hellenic Studies*
Journ. Amer. Orient. Soc.	*Journal of the American Oriental Society*
Journ. Brit. Astron. Assoc.	*Journal of the British Astronomical Association*
Journ. Cuneif. Stud.	*Journal of Cuneiform Studies*
JNES	*Journal of Near Eastern Studies*
JRS	*Journal of Roman Studies*
KR	G. S. Kirk and J. E. Raven, *The Pre-Socratic Philosophers*, 1960
L'Antiq. Class.	*L'Antiquité Classique*
LSJ	H. G. Liddell and G. S. Scott, *A Greek–English Lexicon*, 9th ed., revised by H. S. Jones
MNRAS	*Monthly Notices of the Royal Astronomical Society*
Neugebauer, *ACT*	O. Neugebauer, *Astronomical Cuneiform Texts*, 3 vols., 1955
Neugebauer, *Ex. Sci.*	O. Neugebauer, *The Exact Sciences in Antiquity*, 2nd ed., 1957
OCT	*Oxford Classical Texts*
Philol. Untersuch.	*Philologische Untersuchungen*
Philos.	*Philosophy*
Phron.	*Phronesis*
Proc. Amer. Philos. Soc.	*Proceedings of the American Philosophical Society*
RE	Pauly-Wissowa, *Real-Encyclopädie der classischen Altertumswissenschaft*
Rhein. Mus.	*Rheinisches Museum für Philologie*
Ross, *Arist. Met.*	W. D. Ross, *Aristotle's Metaphysics*, 2 vols., 1924
Ross, *Arist. Phys.*	W. D. Ross. *Aristotle's Physics*, 1936
Van der Waerden, *Anf.*	B. L. van der Waerden, *Die Anfänge der Astronomie*, 1956

1 Part of the material in this chapter has been adapted from my article in *Bull. Inst. of Class. Stud.* No. 11, London, 1964, pp. 43*ff.*

2 Hence a description of the type of instrument used in later Greek astronomy, e.g. by Hipparchus and Ptolemy, has been postponed to the second volume—meanwhile, see *Journ. Brit. Astron. Assoc.* 64, 1954, pp. 72–85

3 Much of this may be found in the elementary handbooks of astronomy written by Geminus in the first century BC (*Isagoge in Phainomena*, ed. Manitius, Teubner, 1898) and by Cleomedes in the first century AD (*Cyclica Theoria*, ed. Ziegler, Teubner, 1891)

4 E.g. by Autolycus of Pitane in his *On Risings and Settings*, ed. Hultsch, Teubner, 1885, the earliest extant Greek astronomical treatise, dating from the last decades of the fourth century BC

5 Adapted from O. Schmidt's paper on Autolycus in *Den 11. skandinaviske matematikerkongress*, Trondheim, 1949, pp. 204–05

6 Ptol., *Geogr.* i, 7, 4

7 *Cf.* Fig. 13, (p. 222) a diagrammatic representation of the terrestrial sphere with a parallel of latitude OO at which the plane of the observer's horizon is HH. Note that the latitude is given by the angle ϕ which by elementary geometry is the same as the height of the north celestial pole above the horizon, but not directly by the angle α, which is the angle that the equatorial plane makes with the horizon and is, in fact, the 'co-latitude' = $90° - \phi$

8 This is the ratio accepted by Hipparchus for Athens (*Comm. in Arat.* i, 3, 6), and gives a latitude about 1° too low – see p. 154

9 A ratio given by Eudoxus – see p. 154

10 On ancient trigonometry, see especially Heath, *Hist. of Greek Maths.*, vol. ii, pp. 257–60, 265–73, 276–86

11 See *GFH*, pp. 159; 162–63

12 See a news story in *The Times* of 14 April 1965, headed 'Error Found in Moon's Orbit'

13 *Cf.* O. Neugebauer, *The Exact Sciences in Antiquity*, 2nd ed., 1957, p. 99, 'Mythological concepts which involve the heavens, deification of Sun, Moon, or Venus cannot be called astronomy if one is not willing to count as hydrodynamics the existence of belief in a storm deity or the personification of a river. Also the denomination of conspicuous stars or constellations does not constitute an astronomical science'

14 See A. Pannekoek, *Hist. of Astron.*, English trans., 1961, ch. 7; Neugebauer, *op. cit.*, ch. 4

15 *Cf. GFH*, p. 14. The idea that astrology is an early, bastard form of astronomy is based on no good evidence at all—*cf.* Neugebauer, *op. cit.*, p. 168

16 Plato's recommendation in the *Laws* of the public worship of sun, moon, and stars (821 *c–d*) was never put into practice

17 See *Hermes* 91, 1963, pp. 60*ff.*

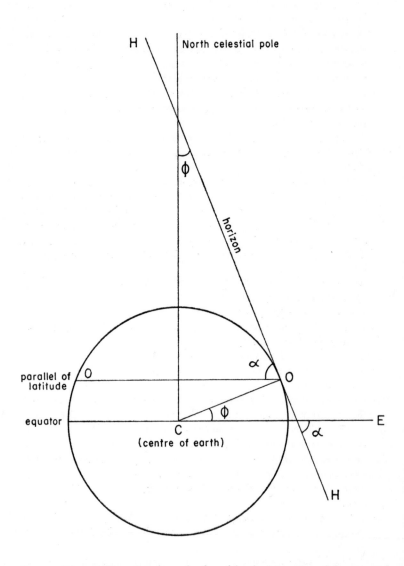

Fig. 13. *Diagram illustrating the angles formed by the intersection of the terrestrial equator and north-south axis with the horizon at north latitude φ*

18 T. B. L. Webster, *From Mycenae to Homer*, 1958, pp. 44–5

19 To suppose that it was regarded as a thick disc is mere anticipation of
later Pre-Socratic views (see p. 58); in reality, no clear idea of the shape
and position of the earth in relation to the heaven and the underworld can
be gained from the poems. At *Il.* viii, 13–16, Zeus threatens to hurl any
disobedient god 'into murky Tartarus, far, far away, where the deepest
chasm is under the earth, and where there are iron gates and a brazen
threshold, as far beneath Hades as the heaven is from earth'. Now obviously
this is the language of poetic imagination, not that of cosmological speculation; but it is also the language of 'mythopoeic' thought, which does
not operate with the same concepts of space, time, cause, effect, subjectivity
and objectivity as were later to be made familiar by the Greek philosophers
and hence absorbed into modern European thought (see H. and H. A.
Frankfort, *Before Philosophy*, Pelican, 1961, ch. 1 and 8). It is therefore
vain to expect the Homeric world picture to exhibit even that imaginative
consistency in detail which Dante displays in his description of earth,
heaven, and hell in the *Divina Commedia* – consistency, a *sine qua non* in
scientific thinking, played a very minor role in pre-scientific (i.e. pre-fifth-
century BC Greek) thought

20 *Cf.* Kirk and Raven, *The Pre-Socratic Philosophers*, 1960, pp. 12–13

21 Which I regret having followed myself in e.g. *JHS* 86, 1966, p. 31

22 The whole passage (x, 82–6) is odd: 'Where herdsman calls to herdsman,
the one driving in his flocks, while the other answers as he drives his out.
There a sleepless man could earn two wages, one tending cattle, and the
other pasturing silvery sheep; for the paths of day and night are close
together.' A sleepless man can *always* earn two wages *anywhere*, especially
as it is made clear that the jobs are different (concerned with cattle and
sheep respectively). One is tempted to suggest the excision of lines 84–5,
i.e. 'There . . . sheep'; then the last line follows closely on the allusion to
a night so short that no sooner has one man driven in his flocks, than
another drives his out. The coalescence of two separate men into one
sleepless man, and the differentiation of cattle and sheep, may have been
an unhappy elaboration introduced by someone who did not understand
the reference to short summer nights in the original lines

23 Eclipses are not mentioned as such, although there is one reference to the
sun's vanishing from the heavens (*Od.* xx, 356–57). This, however, occurs
in the description of the supernatural terror and portents with which
Athene afflicted the minds of the suitors before their impending doom
(*ibid.* pp. 345*ff.*), and in the context can hardly refer to an actual event,
although some of the ancient commentators took it as such (see Stanford
ad loc.); the idea is occasionally revived by modern interpreters, according
to whom a solar eclipse occurred in 1,178 BC with the line of totality
passing through the island of Leucas (assumed to be Homeric Ithaca) – *cf.*
T. L. MacDonald, *Journ. Brit. Astron. Assoc.* 77 (5), 1967, pp. 324*ff.* How

much this is pure coincidence or whether it represents a vestigial memory of an eclipse enshrined in the epic tradition, it is impossible to say. One would, perhaps, have thought that such a rare and impressive phenomenon as a total solar eclipse, if it had actually been experienced, would have merited more than such a brief allusion.

According to Plutarch (*De Fac. in Orb. Lun.* 931e) solar eclipses were mentioned by Mimnermus and Archilochus (both of the seventh century BC), Stesichorus (seventh – sixth century) and Pindar (sixth – fifth century) as well as by Cydias (early-fifth century). Of these only the two mentioned by Archilochus (fr. 74 *Anth. Lyr. Gr.*, ed. Diehl) and Pindar (*Paean* ix, 1–5, fr. 44, Bowra) can be identified with any certainty, the former as the eclipse of 6 April –647 (Jacoby's doubts about this in *CQ* 35, 1941, pp. 97–8 are unnecessary), and the latter that of 30 April –462, being the largest visible at Thebes in Pindar's time (*cf.* J. K. Fotheringham, 'A Solution of Ancient Eclipses of the Sun', *MNRAS* 81, 1921, pp. 104*ff.*; 107; 109)

24 As Finley points out (*World of Odysseus*, 1956, pp. 151–52), all that Helios can do when Odysseus' men kill his cattle (*Od.* xii) is to rush off and complain to Zeus

25 *Cf. Il.* viii, 541, where a particular day is described as 'bringing evil' for the Greeks, and *Od.* i, 283, where rumour 'brings news' for men

26 Normally in the literary sources, unless the context clearly indicates otherwise, as it does at 567, 'rising' (ἐπιτολή, distinguished from the daily rising which is ἀνατολή – *cf.* Geminus, *Isag.* 13) and 'setting' (δύσις) refer to the heliacal rising and cosmical setting, both phenomena occurring before sunrise – *cf.* p. 13

27 The standard works in this field are P. V. Neugebauer, *Tafeln zur astronomische Chronologie*, 1912–25, and *Astron. Chron.*, 1929; the astronomical data in F. K. Ginzel, *Handbuch der math. und techn. Chron.*, 3 Bd., Leipzig, 1906–14 (photo-lith. repr. 1958) are based on Neugebauer's tables, but Ginzel also gives additional tables listing useful data not tabulated by Neugebauer; U. Baehr, *Tafeln zur Behandlung chron. Probleme*, 1955 (Veröffentlichungen des Astron. Rechen-Instit. zu Heidelberg, Nr. 3) repeats some of Neugebauer's tables using more recently worked values for the different constants

28 E.g. in the 'parapegma' texts, see pp. 84*f.*

29 *Cf.* M. P. Nilsson, *Primitive Time-Reckoning*, 1920, pp. 40–1; 49; 56; 64; 89; 115*f.*; 129*f.*; Aratus, *Phain.* 264*f.* and schol. *ad loc.*

30 *Cf.* 586–88, 'Women are most wanton and men most feeble when Sirius parches the head and knees, and the skin is dry because of the heat'; similarly at *Scut. Her.* 397. As in *Il.* xxii, 30–1, it is highly improbable that the figurative language used here implies a belief that the stars actually affected conditions on earth – *cf.* above, p. 95

31 W. Kubitschek, *Grundriss der antiken Zeitrechnung* (in Müller's *Hdbk. d.*

Altert.), 1928, p. 109, fixes it at 28 Dec. for latitude 38°N. about 800 BC

32 Well explained by A. W. Mair in an addendum, entitled 'The Farmer's Year in Hesiod' (especially pp. 130; 142), to his translation of Hesiod's poems (Oxford, 1908)

33 As is stated by the scholiast on Aratus, *Phain.* 254 – *cf.* 172 – and alleged extracts from which are given in Pseudo-Eratosthenes, *Catast.* fr. 1 and 32 (ed. Robert, Berlin, 1878; *cf.* Maass, *Philol. Untersuch.*, Heft 6, 1883, pp. 3–55); *cf.* Diod. Sic. iv, 85; Plato, *Epin.* 990a; Callim. *Ep.* 27

34 *Cf.* my article 'Solstices, Equinoxes, and the Pre-Socratics', *JHS* 86, 1966

35 W. K. C. Guthrie, *History of Greek Philosophy*, vol. ii, 1965, p. 345

36 First ed. 1903; 5th ed. 1934–37 revised and re-edited by W. Kranz in three vols. and reprinted several times since

37 E.g. in *Metaphysics* A – although there is little doubt that his historical notices of earlier doctrines are sometimes coloured by his own pre-conceptions; *cf.* H. Cherniss, *Aristotle's Criticism of Pre-Socratic Philosophy*, 1935

38 *Cf.* CQ 9, 1959, pp. 294–309, especially 298

39 Anthologies of the collected 'opinions' (δόξαι – also ἀρέσκοντα, Latinized as *placita*) of earlier thinkers were best-sellers in late antiquity

40 *Cf.* also Heath, *Arist.*, pp. 2*f.*

41 *Cf. JHS* 86, 1966, pp. 29*f.*

42 G. S. Kirk and J. E. Raven, *The Pre-Socratic Philosophers*, 1960, p. 7 – hereafter referred to as *KR*

43 *Cf.* Aristophanes, *Clouds* 180; *Birds* 1009; Plato, *Theaet.* 174a

44 Both these theorems are attributed to Thales on the authority of Eudemus (Proclus, *Comm. in Eucl. i*, ed. Friedlein, pp. 299, 1; 352, 14). For Thales bringing geometry into Greece from Egypt, see the 'Eudemian summary' in Proclus (*op. cit.*, p. 65, 7; Wehrli, *Die Schule des Arist.*, Heft 8 – Eudemos von Rhodos, fr. 133)

45 Which would require, as well as an accurate knowledge of solar and lunar cycles and the moon's deviations in latitude from the ecliptic, an understanding of the concept of geographical latitude in order to be able to predict the totality of the eclipse in a given region. Such an advanced level of knowledge was not even reached by Babylonian astronomy of the Seleucid period (the last three centuries BC), much less that of the sixth century BC

46 *Cf.* the creation myths of the ancient Egyptians and Babylonians and of *Genesis* (*KR*, ch. 1, especially pp. 33*f.*). Recent speculation has attributed a similar type of cosmogony, with the personified figures of Πόρος and Τέκμωρ (apparently 'Contrivance' and 'Differentiation') to the seventh-century BC Spartan poet, Alcman – see M. L. West, *CQ* 13, 1963, pp. 154–56, and *CQ* 17, 1967, pp. 1–15

47 The same word is used by Socrates (*Phaedo* 109a2) of the heavens (οὐρανός) in a similar context – see p. 95

15

48 *Cf.* the waters of the firmament in *Genesis* 1

49 See especially C. H. Kahn, *Anaximander and the Origins of Greek Cosmology*, Columbia Univ. Press, 1960, *passim*

50 Compare the accounts of Anaximander's supposed 'system' given by Heath, *Arist.*, 1913, ch. 4; J. S. Morrison, *JHS* 75, 1955, pp. 62–3; *KR*, pp. 134–37; Kahn, *op. cit.*, pp. 87*ff.*; Guthrie, vol. i, pp. 93–9; O'Brien, *CQ* 61, 1967, pp. 423*ff.*

51 It is very improbable that the planets were yet differentiated (*cf. JHS* 86, 1966, p. 30), and, if they were not, occultations of stars by them would presumably not be noticed, so that Anaximander's alleged order of the celestial bodies might not seem so crass; but occultations of stars by the moon are relatively frequent and more easily observable

52 Strangely, no modern booster of the Pre-Socratics as 'super-scientists' has, to my knowledge, seized on this as 'convincing evidence' for their understanding of the principle of the hovercraft . . . No doubt this will come in time – we have already had it suggested that Wegener's theory of continental drift was anticipated by Thales' earth floating on water! (K. R. Popper, 'Back to the Pre-Socratics', *Proceedings of the Aristotelian Society*, Oct. 1958, p. 3). For a sensible rebuttal of such fancies see G. S. Kirk, 'Popper on Science and the Pre-Socratics', *Mind* 69, 1960, p. 328

53 The dates of the Pre-Socratics are largely guesswork, although some guesses are more informed than others

54 This type of contradiction is frequent in the tertiary sources, and serves to demonstrate why they are so untrustworthy as independent evidence

55 Heath (*Arist.*, p. 42–3) also compares another passage in Aëtius (ii, 23, 1 = DK 13 A15), which speaks of 'the stars making their turnings by being thrust off course' by wind (ἐξωθούμενα τὰ ἄστρα τὰς τροπὰς ποιεῖσθαι). This is another notion that we shall meet again in connection with later Pre-Socratics

56 *DK* 21 A32; A33; A37–41; A45

57 His is the famous dictum πάντα ῥεῖ (*cf.* Plato, *Theaet.* 181*a*), and the remark that you cannot step into the same river twice (*DK* 22 B31)

58 His is the first extant use of the actual word κόσμος in its philosophical sense of 'world-order', i.e. practically equivalent to 'universe' – *cf.* J. Kerschensteiner, *Kosmos*, Munich, 1962 (Zetemata, Heft 30)

59 J. H. Morrison in *JHS* 75, 1955, accepting every scrap of evidence from whatever source, has attempted a reconstruction which is based on attributions to both Anaximander and Parmenides of astronomical knowledge (e.g. of the ecliptic) which it is highly improbable that they possessed (see my article in *JHS* 86, 1966)

60 This from Strabo via Posidonius, A44*a* – it is, of course, completely wrong, as the theory of zones is a gross anachronism for Parmenides' time: *cf.* my *GFH*, pp. 23–6

61 ἀλλότριον φῶς, properly 'a light belonging to someone or something

else' – the phrase is evidently a pun on the Homeric ἀλλότριος φώς, 'a foreign man', *Il.* v, 214; *Od.* xviii, 219

62 *Cf.* fr. 115, 11, αἰθέρος . . . δίναις and the αἰθέριος δῖνος of Aristophanes, *Clouds* 380

63 ἀλισθείς – not necessarily 'globulated', as some translators have it

64 Unfortunately, the text of Plutarch, *De Fac. in Orb. Lun.* 929c (which provides the fragment) is uncertain; for the MS ἀπεσκεύασε δὲ οἱ αὐγάς, ἔστε αἶαν καθύπερθεν, Diels conjectured ἀπεστέγασεν . . . ἔστ' ἂν ἴῃ καθύπερθεν, followed by *KR* who translate (p. 334), 'But she kept off the sun's rays, so long as it was passing over above her, and cast a shadow over as much of the earth as was the breadth of the pale-faced moon' (*cf.* Guthrie, vol. ii, p. 197). Yet the verb ἀποστεγάζω, here translated 'kept off', is used again in fr. 100, 14 where *KR* translate it 'uncover' (the meaning given by *LSJ, s.v.*), which is also how Diels translates his own conjecture (*DK*, vol. i, p. 330, '*Der Mond* deckte ihr (*der Sonne*) die Strahlen ab'). Cherniss (*Moralia*, vol. xii, Loeb, *ad loc.*) retains the MS reading, but prints ἀπεσκέδασεν for ἀπεσκεύασε, and translates, 'His beams she put to flight . . . From heaven above as far as to the earth, Whereof such breadth as had the bright-eyed moon She cast in shade'

65 Compare the famous clepsydra simile in B100 (also from Aristotle – *De Resp.* 7, 473b9), designed to illustrate the corporeality of air by the behaviour of water in a double-ended vessel when one end is blocked up. This is *not*, as sometimes claimed (e.g. by Burnet, *Early Greek Philosophy*, p. 27), an instance of the use of the experimental method in ancient science – see *KR*, p. 342; Guthrie, vol. ii, pp. 225–26

66 Unless Aristotle misunderstood the point of Empedocles' analogy. For a discussion of various views on this passage, see Guthrie, vol. ii, p. 198–99

67 J. Longrigg in *CQ* 15, 1965, has an ingenious notion that this idea might well have been 'derived from his observation of the formation of salt under the action of the sun's heat'! (p. 251)

68 Aristotle obviously had a similar feeling when he contrasts the sobriety of Anaxagoras' views with the random opinions of his predecessors (*Met.* A 984b17)

69 *Cf.* Xenophon (*Mem.* iv, 7), who makes Socrates assert that Anaxagoras must have been out of his mind (παρεφρόνησεν) to have said such things, and makes him refute them by some painfully naive arguments

70 The scientific ideas that Aristophanes pokes fun at and attributes to Socrates and his 'thinking-shop' in the *Clouds*, are mainly those of Anaxagoras

71 His word for the primal movement is περιχώρησις, 'rotation', instead of δίνη

72 Diog. Laert. ii, 12 (*cf.* Diod. Sic. xii, 39). Plutarch (*Per.* 32) connects Anaxagoras' impeachment with a decree proposed by one Diopeithes, *c.* 431 BC, and aimed at 'those who do not believe in the gods or who teach doctrines about celestial matters'. This, together with Plutarch's

exaggerated description of the hostility of the Athenian public to new astronomical ideas (*Nic.* 23 – see my forthcoming review in *CR* of T. W. Africa's *Science and the State in Greece and Rome*, 1968) has led some scholars (e.g. J. B. Bury, *CAH*, vol. v, p. 383) to suppose that the study of astronomy was officially forbidden at Athens during the latter part of the fifth-century BC. Nothing could be further from the truth. As we shall see, this was the time when Greek mathematical astronomy had its beginnings, with the collation and codification of astronomical and meteorological phenomena in the 'parapegmata', the diffusion of Pythagorean ideas on the sphericity of the earth and the circular orbits of celestial bodies, the discovery of the obliquity of the ecliptic by Oinopides, and the activities of Meton and Euctemon in measuring solstices and equinoxes and improving the calendar.

The circumstances of Anaxagoras' impeachment are by no means clear – see the discussions by Taylor (*CQ* 11, 1917, pp. 81–7), Davison (*CQ* n.s. 3, 1953, pp. 42–5), and Guthrie (vol. ii, pp. 322–23); but it seems most likely that it was really Pericles who was being attacked through his old master, Anaxagoras. As a former Persian subject, Anaxagoras also came under suspicion of 'medism'. The fact that he advocated unorthodox opinions about the celestial bodies may have helped the prosecution, but it is highly improbable that this alone would have been sufficient to condemn him – just as a similar charge did not suffice in the case of Socrates, who was condemned mainly because of his anti-democratic connections and sentiments

73 *Cf.* K. R. Popper, 'Back to the Pre-Socratics', *Proceedings of the Aristotelian Society*, Oct. 1958, p. 3, '. . . but most of them [the ideas of the Pre-Socratics], and the best of them, have nothing to do with observation'. Kirk notes the 'superficial glance which was all that many Pre-Socratics seem to have considered necessary' (*Mind* 69, 1960, p. 329)

74 The mathematical relationship expressed by which is now known to have been familiar to the Babylonian mathematicians at least a thousand years before Pythagoras

75 This, as Heath points out (*Arist.*, p. 99), entails one complete rotation of the earth on its axis as it completes one circuit of the central fire

76 This, however, is a stock title which the doxographers attribute to practically every single Pre-Socratic – *cf. KR*, p. 101

77 On the nomenclature of the planets see F. Cumont, 'Les noms des planètes et l'astrolatrie chez les Grecs', *L'Antiq. Class.* 4, 1935, pp. 5–43

78 Unless, indeed, the moon was supposed to reflect the light of the central fire and not that of the sun. Apparently, some Pythagoreans realized that the moon shone by reflected sunlight (Diog. Laert. viii, 27), but others still spoke of its own fire being kindled, spreading to the state of full moon, and then being gradually extinguished (Aëtius, *DK* 58 B36)

79 *Cf.* the use of the word ἀνταίρειν in a geographical sense – *GFH*, p. 125

80 The Greek is οἱ δὲ γνησιώτερον αὐτῶν μετασχόντες πῦρ μὲν ἐν τῷ μέσῳ λέγουσι τὴν δημιουργικὴν δύναμιν τὴν ἐκ μέσου πᾶσαν τὴν γῆν ζῳογονοῦσαν καὶ τὸ ἀπεψυγμένον αὐτῆς ἀναθαλποῦσαν. Surprisingly, this has been interpreted as referring to another brand of Pythagoreanism which believed in a fiery core *at the centre of the earth*, itself situated in the middle of the universe – as though the words ἐν τῷ μέσῳ and ἐκ μέσου referred to the earth's centre and not the centre of the cosmos (H. Richardson, *CQ* 20, 1926, pp. 118–21; Guthrie, vol. i, pp. 290f.). Not only does the context show this interpretation to be impossible, but there is no good evidence for the idea of a fiery core to the earth in Pythagorean thought. Guthrie can only point to a similar idea in Empedocles and assume that this might have formed part of early Pythagoreanism, and the single piece of evidence for a geocentric universe in this school (which entirely contradicts what we are told by Aristotle who, after all, did write a book on the Pythagoreans) is the muddled account of Diogenes Laertius (viii, 25 *ad fin.*), ostensibly reporting Alexander Polyhistor, which does not mention the Philolaic scheme at all, but attributes a 'spherical cosmos possessing soul and intellect and surrounding the central earth that is also spherical and inhabited all round' to 'Pythagorean writings'. There is obvious confusion with Stoic ideas here

81 This is implicit in Simplicius' words, (pp. 511–12 ed. Heiberg, ἡ δὲ ἀντίχθων κινουμένη περὶ τὸ μέσον καὶ ἑπομένη τῇ γῇ ταύτῃ οὐχ ὁρᾶται ὑφ'ἡμῶν διὰ τὸ ἐπιπροσθεῖν ἡμῖν ἀεὶ τὸ τῆς γῆς σῶμα)

82 As Dreyer sensibly remarks (*Hist. of Astron.*, Dover reprint, 1953, p. 48), it would require an observer at the centre of motion itself to differentiate between a motionless outer heaven and an earth revolving in exactly twenty-four hours, and a slowly rotating outer heaven with an earth revolving in a slightly shorter period

83 B. L. Van der Waerden, *Die Astronomie der Pythagoreer*, Amsterdam, 1951, p. 54. The book gives a totally misleading idea of its subject matter, and is based largely on what the author thinks Pythagorean astronomy *ought* to have been; thus, to explain the Philolaic system Van der Waerden makes up his own 'einzige Lösung' (p. 60) according to which the order of the orbits from the centre is sun, Mercury, Venus, earth (with moon encircling it), Mars, Jupiter, and Saturn – an order for which, needless to say, there is not a shred of evidence and which contradicts the little we are told about the scheme

84 *Arist.*, p. 100 – but his translation of the passage '. . . that it revolves round the fire in an oblique circle in the same way as the sun and moon' (p. 97) does not square with such an interpretation (which Dreyer also seems to accept – *op. cit.*, p. 45). As far back as 1810, A. Boeckh evidently took the same view (*Gesammelte Kleine Schriften*, Bd. 3, Leipzig, 1866, pp. 266; 276f.)

85 This is obviously the motivation of Boeckh, Heath, Dreyer, and Van der

Waerden as well; but it is surely not justifiable in the face of the bulk of the evidence

86 *Cf.* Guthrie, *Aristotle On the Heavens*, Loeb, 1960, pp. 136–37

87 *Cf.* Van der Waerden, *Anfänge der Astronomie*, 1966, pp. 49–50. On the influence of Babylonian astronomy on Greek, see above, pp. 163*ff.*

88 Hence Diels omits it from *DK* 58 – *cf. Dox. Gr.*, Prolegomena, p. 69

89 *Cf. Timaeus* 39*d*

90 *I precursori di Copernico nell'antichita*, Milan, 1873, p. 8

91 Heath, *Hist. of Greek Maths.*, vol. i, pp. 246–49

92 *Odes* i, 28, 1–6

93 Hence the unnecessary doubts that have been raised as to whether Plato regarded the earth as spherical; but Socrates in the *Phaedo* opposes στρογγύλος to πλάτυς, 'flat' (97*d* – see above, p. 94), which is the word specifically connected with the Anaxagorean disc-shaped earth (see p. 58), and so in this case there is nothing to prevent στρογγύλος meaning 'spherical' – and the rest of Plato's astronomical passages undoubtedly imply a spherical earth (see p. 98)

94 Or, perhaps more accurately, there was a common stock of astronomical notions ascribed to earlier thinkers from which all tertiary sources draw

95 The notion that the celestial bodies are nourished by exhalations from the earth (probably on the analogy of burnt offerings to the gods) is attributed by Aëtius to both Heraclitus and Philolaus (*DK* 22 A11; 44 A18)

96 sed nec numerum illarum posuit nec nomina, nondum comprehensis quinque siderum cursibus

97 See A. Rehm's article in *RE*, Bd. 18 (4), 1949, col. 1299*f.* H. Diels (*Antike Technik*, 3rd edition, 1924, Tafel 1, pp. 6–7) gives an illustration of one fragment

98 See the detailed study by Rehm, 'Parapegmastudien', *Abhandl. d. Bayerischen Akad. d. Wiss.*, phil.-hist. Abt., neue Folge, Heft 19, 1941, pp. 5*ff.*

99 Ed. Manitius, Teubner, 1898, pp. 210–33; Manitius shows that the calendar actually belongs to a period about one hundred years earlier than the main text which is dated to *c.*70 BC

100 Ed. Heiberg, vol. ii, 1907, *Claudii Ptolemaei Opera Astronomica Minora*, Teubner, pp. 1–67

101 Other Greek calendars are published in *Sitz.-Berichte Akad. Heidelb.*, phil.-hist. Kl. I (1910), II (1911), IIIA (1913) – this contains a conjectural restoration by Rehm of Euctemon's parapegma (on whom see below) – IV (1914), and V (1920)

102 ἐπισημασίαι, 'signs *or* indications of weather' – ἐπισημαίνειν is used of marking a change in the weather; *cf.* Rehm's article 'Episemasiai' in *RE*, Suppl. Bd. 7, 1940, cols. 175–98

103 O. Neugebauer, *The Exact Sciences in Antiquity*, 2nd ed., 1957, ch. 5

104 M. P. Nilsson, *Primitive Time-Reckoning*, p. 343; *Die Entstehung . . . griech. Kal.*, 1962, pp. 31*f.*; 56

NOTES 231

105 Pliny, *Nat. Hist.* ii, 31 – on this see *JHS* 86, 1966, pp. 26–7; 30
106 E.g. J. K. Fotheringham, in *JHS* 39, 1919, pp. 164–84; refuted in detail by
 E. J. Webb in *JHS* 41, 1921, pp. 70–85; Fotheringham's reply in *JHS* 45,
 1925, pp. 78f., based on incorrect ideas of early Babylonian astronomy
 (including the so-called Babylonian 'saros' – on which see Neugebauer,
 Ex. Sci., pp. 141–42), is not convincing
107 *Isag.* ch. 8 – Geminus connects Euctemon but not Meton with this cycle
 (§ 50), but Meton's share in the discovery is amply attested by other
 sources (*cf.* Heath, *Arist.*, p. 293) and it is possible that his name may have
 dropped out of the text – see Manitius *ad loc.*
108 This figure is expressly mentioned by Hipparchus as the estimate of Meton
 and Euctemon – *Almag.* iii, 1, p. 207, 9 ed. Heiberg
109 Ed. F. Blass, 1887 – it is not, of course, by Eudoxus himself, but may be
 a student's exercise based partly on Eudoxan data, but containing many
 errors and with later material added
110 Quoted by Heath, *Arist.*, p. 132
111 See the calendar compiled by W. H. S. Jones (Loeb *Hippocrates*, vol. i, pp.
 67–8)
112 *Cf.* R. Joly, *Hippocrate: Du Régime*, Budé, 1967, pp. xiv–xvi
113 On the whole subject see B. L. van der Waerden in *JHS* 80, 1960, pp.
 168–80; W. K. Pritchett in *Historia*, Bd. 13, Heft 1, 1964, pp. 21–36
114 He suggested a method of squaring the circle by successively inscribing
 regular polygons with double the number of sides of the previous one – see
 Heath, *Hist. of Greek Maths.*, vol. i, p. 222
115 Heath, *Hist. of Greek Maths.*, vol. i, pp. 23; 225
116 *Cf.* A. Schlachter, *Der Globus*, ed. F. Gisinger, 1927 (Stoicheia, Heft 8),
 pp. 174f. The author's attribution of knowledge of the celestial sphere and
 advanced astronomical ideas to Anaximander and Parmenides should be
 disregarded
117 With the single exception that a fifth-century AD writer on rhetoric,
 Sopater, attributes to him the opinion that the sun is a molten lump (*DK*
 82 B31)
118 E.g. A. E. Taylor (*A Commentary on Plato's Timaeus*, Oxford, 1928) was
 firmly convinced that the whole world-system expounded in that dialogue
 was not Platonic at all, but fifth-century Pythagorean – a 'Taylorian
 heresy' that has been sufficiently refuted by F. M. Cornford in *Plato's
 Cosmology*, 1937, pp. ixf. On the other hand, the extent to which what
 are generally regarded as typically 'Platonic' doctrines (e.g. the theory
 of Ideas, the immortality of the soul, and the recollection of knowledge)
 are the creations of Socrates or of Plato himself remains a question which
 is as insoluble as ever (*cf.* C. J. de Vogel, *Phronesis* 1, 1955, pp. 26–35)
119 Thus the description of the spindle of Necessity with its eight whorls
 (*Rep.* x, 616b–617b) forms part of the myth of Er which is avowedly only
 a tale (*cf.* the beginning, 614b, ὅς ποτε . . . and the end, 621b, καὶ οὕτως, ὦ

Γλαύκων, μῦθος ἐσώθη . . .), and the whole of Timaeus' account of the genesis of the universe (including, therefore, the astronomical section, *Tim.* 36b–40d) is no more than a 'likely tale' with which mere mortals must be satisfied (*id.* 29d)

120 One is irresistibly reminded of Fitzgerald's Omar Khayam,

> *Myself when young did eagerly frequent*
> *Doctor and Saint, and heard great argument*
> *About it and about: but evermore*
> *Came out by the same door where in I went*
>
> (Stanza 27)

121 Because both Simmias and Cebes had consorted with the Pythagorean Philolaus (61d6–7), Simmias' assent has been used as evidence (somewhat illogically) for a non-Philolaic, presumedly early, version of Pythagorean cosmology, in which the spherical earth and not the central fire (note 80) was placed at the centre of the universe (*cf.* Burnet, *EGP*, p. 297 note 3). This is at complete variance with what Aristotle tells us about Pythagorean doctrines, as is also the only other evidence for this belief, namely, the garbled account of Diogenes Laertius (Diels, *Vorsokr.* p. 58 B1a6); when it comes to choosing between the latter and Aristotle, there is little doubt as to which to follow. Moreover, Simmias' assent need only refer to the statement that the earth needs no support, which is equally apposite for the planetary earth in the Philolaic system itself; the conditional form in which Socrates states the central position and circularity of the earth (despite the fact that there can be little doubt that this is Plato's own view – see below) may have been purposely chosen so as to command Simmias' assent to the proposition that the earth needs nothing to support it. Socrates might have thought, 'I know that you, as a Pythagorean, do not agree that the earth is at the centre of the universe, but grant me for the sake of argument that it is, then you will certainly agree that it needs no support, won't you?', and Simmias, of course, does

122 This is probably a reference to the dodecahedron, which can be constructed from twelve regular pentagons; its volume approaches closely that of the circumscribed sphere, and if the pieces are made of flexible material stitched together (as here) and filled out, a sphere would result – see Burnet *ad loc.* In the *Timaeus* (55d–56b) four of the five regular solids are appropriated to the four elements (earth = cube, fire = pyramid or tetrahedron, air = octahedron, and water = icosahedron), but the fifth, the dodecahedron, which cannot be constructed from Plato's basic right-angled triangles, is merely stated somewhat vaguely to be used by the god 'for the whole' (ἐπὶ τὸ πᾶν, 55c) – very possibly with this passage of the *Phaedo* in mind. It is highly improbable that any allusion to the twelve signs of the zodiac is intended, which is an entirely different concept, and still less likely that an astronomical system of mapping the sky into twelve zones

is referred to, for which there is no evidence at all (see Cornford, *Plato's Cosmology*, 1937, p. 219)

123 J. A. Stewart, *The Myths of Plato*, ed. G. R. Levy, 1960, p. 119

124 The situation is similar in modern science fiction; it is the writers who operate on a plausible basis of modern scientific concepts (of course, imaginatively employed and extended) who succeed best in this genre

125 Aristotle, characteristically, fails to realize this, and subjects Plato's fanciful picture of the earth's subterranean waters (111*d*–112*e*) to a solemn, scientific criticism (*Meteor*. ii, 2, 355*b*32*ff*.); *cf.* also Friedlander, *Plato*, vol. i (English trans. 1958), p. 267

126 No ancient commentator ever had any doubts as to this, but some modern scholars have exercised their ingenuity in manufacturing them; e.g. T. G. Rosemeyer, *CQ* 6, 1956, pp. 193–97, and *Phronesis* 4, 1959, pp. 71–2; J. S. Morrison, *Phron.* 4, 1959, pp. 101–19

127 Presumably, this implies indirect observation of the sun's image on a liquid surface, not direct observation through, e.g., a liquid-filled flask, since the words φάντασμα and εἰκών would hardly be appropriate for the latter method. Since Socrates is still talking about the time when he was a young man, unless Plato is here committing an anachronism (which is by no means improbable), this would seem to indicate that the observation of eclipses was a well-known proceedure in the middle of the fifth-century BC

128 J. Adam, *The Republic of Plato*, 2nd ed., vol. ii, p. 67

129 For which see the monumental bibliography compiled by H. Cherniss in *Lustrum* 4–5, 1959–60

130 Some of the more percipient treatments of this and other topics in Platonic philosophy are to be found in the collection of articles entitled *Studies in Plato's Metaphysics*, ed. R. E. Allen, 1965

131 *Met.* A 6, especially 987*b*14*ff.*; for the other passages imputing this doctrine to Plato, see W. D. Ross, *Aristotle's Metaphysics*, vol. i, 1924, p. 166 *ad loc.*

132 *Cf.* A. Wedberg, *Plato's Philosophy of Mathematics*, 1955, who upholds Aristotle's interpretation; for weighty arguments against it, see H. Cherniss, *Aristotle's Criticism of Plato and the Academy*, vol. i, 1944, pp. 226*f.*; 244; 289*f.* and notes

133 *Cf.* Wedberg, *op. cit.*, p. 32, 'Between extreme nominalism, which denies the existence of any abstract entities, and an extreme realism, which accepts more or less wholesale all significant expressions as designating entities (abstract if not concrete), there is an entire spectrum of possible positions . . . Plato did never clearly define where along this spectrum he took his stand'

134 Socrates is somewhat unfair here; there is no reason why Glaucon's remark should be taken absolutely literally – he might himself have been thinking of astronomy in a metaphysical way – particularly in view of Socrates' own phrases earlier where he speaks of 'leading the soul upwards'

(ἄνω ποι ἄγει τὴν ψυχήν, 525d) and 'making it see the Idea of Good' τὸ ποιεῖν κατιδεῖν τὴν τοῦ ἀγαθοῦ ἰδέαν, 526e)

135 There is an obvious allusion to the comic caricature of 'Socratic science' in Aristophanes *Clouds* 188*ff.*; *cf.* Plato *Apol.* 19c

136 E.g. H. S. Williams, *Hist. of Sci.*, 5 vols., 1904–10, vol. i, pp. 180*ff.* (Plato's point of view 'essentially non-scientific'); W. C. D. Dampier-Whetham, *Hist. of Sci.*, 1929, p. 28 (Plato's science 'fantastic . . . he roundly condemned experiment'); F. K. Richtmyer, *Introd. to Mod. Phys.*, 2nd ed., 1934, p. 9; C. Singer, *Short Hist. of Sci.*, repr., 1943, pp. 34*f.* (but Singer does see some redeeming features in Plato, especially for mathematics and astronomy); B. Russell, *Hist of West. Philos.*, 1945, p. 165 (the *Timaeus* is 'silly'); J. Jeans, *The Growth of Physical Sci.*, 1947, pp. 47*ff.* (Plato, a 'major disaster' for physics); J. O. Thomson, *Hist. of Anc. Geogr.*, 1948, p. 101; B. Farrington, *Greek Sci.*, Pelican, 1953, p. 120

137 E.g. J. Adam, *The Republic of Plato*, 2nd ed., 1965, Appendices I and II to Book vii

138 The translation is my own but obviously owes much to other versions, of which P. Shorey's (in the Loeb *Plato*) probably comes closest to it. The Greek text (about which there are no serious ms doubts) is as follows (529c–d): ταῦτα μὲν τὰ ἐν τῷ οὐρανῷ ποικίλματα, ἐπείπερ ἐν ὁρατῷ πεποίκιλται, κάλλιστα μὲν ἡγεῖσθαι καὶ ἀκριβέστατα τῶν τοιούτων ἔχειν, τῶν δὲ ἀληθινῶν πολὺ ἐνδεῖν, ἃς τὸ ὂν τάχος καὶ ἡ οὖσα βραδυτὴς ἐν τῷ ἀληθινῷ ἀριθμῷ καὶ πᾶσι τοῖς ἀληθέσι σχήμασι φορᾶς τε πρὸς ἄλληλα φέρεται καὶ τὰ ἐνόντα φέρει. The infinitive ἡγεῖσθαι depends on δεῖν understood from Glaucon's question; τῶν ἀληθινῶν I take to mean the true realities compared and contrasted with τῶν τοιούτων which are the things of the visible world; ἃς . . . φοράς is adverbial accusative serving to define both πολὺ ἐνδεῖν and φέρεται, and τὰ ἐνόντα must, in my view, mean the actual celestial bodies which are carried round in the orbits that Plato hypostatizes as 'real speed' and 'real slowness'. For other interpretations of this passage, see Adam, App. X to Book vii

139 προβλήμασιν ἄρα, ἦν δ'ἐγώ, χρώμενοι ὥσπερ γεωμετρίαν οὕτω καὶ ἀστρονομίαν μέτιμεν, τὰ δ'ἐν τῷ οὐρανῷ ἐάσομεν – note the last word; Plato does not use here the phrase χαίρειν ἐᾶν, 'to renounce' or 'dismiss altogether' (frequent elsewhere in Plato, e.g. *Phaedo* 63e; *Prot.* 348a; *Phileb.* 59b), as he might well have done had he meant to imply that observation could be dispensed with entirely, but the simple verb meaning 'to let be' or 'leave on one side', evidently for the time being while the astronomer concentrates on the mathematical side of his subject

140 Many interpreters of the Line have been led astray by a failure to appreciate the twofold purpose of the simile, and, by trying to force Plato's thought into a rigidly consistent scheme, have rendered it a good deal more obscure. For an eminently sensible treatment of the Line, see J. L. Stocks, *CQ* 5, 1911, 73–88 (p. 76, 'The Line is not a progression'; p. 78, 'Thus mathe-

matics becomes, what Plato in his theory of education requires it to be, the stimulus to philosophy'); *cf.* F. M. Cornford, *Mind* 41, 1932, pp. 37–52 and 173–190 ('The distinction of objects [in the Line] is a matter of expediency in teaching')

141 As E. Dönt well remarks, it is not the observable phenomena of the heavens that Plato regards as unchanging and eternal, but 'unwandelbar sind nur die Gesetze, die dem Wunderbau des Himmels zugrundeliegen' ('Platons Spätphilosophie und die Akademie', *Österreichische Akademie der Wissenschaften, Sitzungsberichte*, phil.-hist. Klasse, Bd. 251, Abh. 3, Wien, 1967, p. 65)

142 *Cf.* the similarly hyperbolical statement that the man out of the Cave will be able to 'look on the actual sun itself' (516b – quoted above)

143 Well represented by Neugebauer (*Ex. Sci.*, 2nd ed., p. 152): 'His advice to the astronomers to replace observations by speculation would have destroyed one of the most important contributions of the Greeks to the exact sciences'

144 His standpoint in this respect is consonant with his hostility to the views of Democritus, who composed a 'parapegma' or at least made observations for one (see above)

145 There is no evidence anywhere in the extant works of Plato that he ever conceived of anything other than the earth as lying at the centre of the universe; nor does Aristotle attribute any such notion to him. The well-known story found in two passages of Plutarch (*Quaest. Plat.* viii, 1 and *Numa* 11 – the former by hearsay, φασί, but the latter allegedly on the authority of Theophrastus) that, towards the end of his life, Plato repented of having given the earth this central position, which should be occupied by something better, may safely be dismissed as a speculative piece of gossip or a misunderstanding of the sources.

146 This, of course, is not the first time this has been pointed out – *cf.* A. N. Whitehead, *The Concept of Nature*, 1920, p. 18 ('Plato's guesses read much more fantastically than Aristotle's systematic analysis; but in some ways they are more valuable. The main outline of his ideas is comparable with that of modern science'); *Sci. and the Modern World*, 1926, pp. 42–3; P. Shorey, 'Platonism and the History of Science', *Proc. Amer. Philos. Soc.* 66, 1927, pp. 159–82; A. Rey, *La sci. dans l' antiq.*, 6 vols., 1930–48, vol. iii, 282*ff.*; G. C. Field, 'Plato and Natural Sci.' *Philos.* 8, 1933, pp. 131–41; A. Momigliano (reviewing Farrington's *Sci. and Polit. in the Anc. World*, 1939) in *JRS* 31, 1941, pp. 149*ff.*; J. E. Boodin, 'The discovery of form', *J. Hist. Id.* 4, 1943, p. 191 ('. . . Platonic intuition of form and measurement everywhere'); W. Heisenberg, *Philos. Probs. of Nucl. Sci.*, trans. F. C. Hayes, 1952, pp. 33*f.* (Plato's stress on mathematical laws of nature underlying natural phenomena); p. 57; and on the whole subject, G. E. R. Lloyd, 'Plato as a Natural Scientist', *JHS* 88, 1968, pp. 78–92

147 Plato seems here to have taken over the semi-mystical Pythagorean

conception of numbers as the basic 'stuff' out of which the universe is composed (see pp. 64f.). There is obviously less justification in the case of harmonics for a purely theoretical treatment than in astronomy, and it would seem that Plato in this instance allowed himself to be unduly influenced by Pythagorean notions, in the interests of emphasizing his cherished conviction that mathematical harmony underlies the whole phenomenal world – *cf.* the number mysticism of *Rep.* viii, 546b–c, the notorious 'Platonic number', which has nothing to do with astronomy

148 Taking ἄνωθεν with καθορᾶν (*cf.* 616d, κύκλους ἄνωθεν τὰ χείλη φαίνοντας and *Phaedo* 110b, εἴ τις ἄνωθεν θεῷτο) – others take it with τεταμένον 'stretched from above' (*cf.* Adam, *ad loc.*)

149 Adams translates περιφοράς as 'revolving spheres', which is misleading as there is no mention of spheres in the description at all; περιφορά means the particular form of movement appropriate to a circle, i.e. revolution or rotation (*cf.* Arist., *De Anima.* i, 3, 13, νοῦ μὲν γὰρ κίνησις νόησις, κύκλου δὲ περιφορά)

150 Cornford's picture of a 'nest of hemispherical bowls' (*Plato's Cosmology*, p. 75; *The Republic of Plato*, p. 350) is an unwarranted inference from the actual Greek text which mentions neither spheres nor hemispheres. We are told simply that the whorls fit inside each other καθάπερ οἱ κάδοι οἱ εἰς ἀλλήλους ἁρμόττοντες. Now the κάδος in shape is generally taken to resemble the amphora, having two handles and a neck narrower than its belly (*cf.* Amyx in *Hesperia* 27, 1958, pp. 186–90); this, of course, makes it impossible for such κάδοι to fit into one another. It seems likely that here Plato is using κάδος in its other sense of a measure of volume (see *LSJ s.v.*); two such standard measures, in the form of bronze cylinders, one fitting inside the other, have actually been found and are dated to c.400 BC (*The Athenian Agora*, vol. x, 1964, Pl. 14, DM 42 and 43). The notion of hemispherical bowls is in any case excluded by the description of the outer whorl as being 'hollow and scooped out all through' (κοίλῳ καὶ ἐξεγλυμμένῳ διαμπερές) and the spindle's shaft as being 'driven right through' (διαμπερὲς ἐληλάσθαι); rather, we should think of thick, cylindrical discs, fitting closely inside each other, but each able to rotate round the shaft. It is possible that Plato conceived of the discs as tapering slightly, so that the whole shape would be that of a truncated cone; Strabo uses the word σπόνδυλος to describe that segment of the terrestrial globe lying between the equator and the Arctic Circle (Str. ii, 5, 6, C 113)

151 Both interpretations go back to the ancient commentators: the former to Proclus (*in Remp.* ii, p. 130, 4 ed. Kroll), upheld by A. Boeckh, *Kleine Schriften*, 1866, vol. iii, pp. 298ff., and recently revived by J. S. Morrison in *JHS* 75, 1955, pp. 66–7 (who does not mention Boeckh); and the latter to Theon of Smyrna (*Ad Leg. Plat. Util.*, p. 143 ed. Hiller), preferred by Adam (vol. ii, pp. 446–47)

152 διὰ παντὸς τοῦ οὐρανοῦ καὶ γῆς τεταμένον – Boeckh (duly followed by

Morrison) unconvincingly takes διά here in the sense of 'over', despite
the clear parallel in *Tim.* 40b τὸν διὰ παντὸς πόλον τεταμένον (*cf.* Adam,
vol. ii, App. VI, p. 471)

153 οἷον τὰ ὑποζώματα τῶν τριήρων – again there is doubt as to the meaning;
the reference might be to ropes passed horizontally round a ship from stem
to stern, or to ropes going under the keel and up the sides in a vertical
plane, both types serving to strengthen the hull (see Adam *ad loc.*)

154 *Cf.* Adam *ad loc.*; Cornford, *Tim.*, p. 88 and *Rep.*, p. 354; Heath, *Greek
Astron.*, p. 48;. Shorey in the Loeb *Rep.* translates 'with returns upon
itself', which is hardly illuminating

155 If Zeus is the sphere of the fixed stars, are we to imagine the other gods
as standing and being carried round on his back?

156 R. Hackforth, *Plato's Phaedrus*, 1952, pp. 72–3; but the reason he gives
(p. 74, '. . . it seems an insuperable objection that the planets of Greek
astronomy did not have hosts of satellites') is irrelevant – there is no question
of planetary satellites, but of a multiplicity of gods and daemons which is
well attested for Plato (*cf. Polit.* 272e; *Tim.* 40d; 41a; *Rep.* 617e; 620d;
Laws 899b; *Epin.* 948d)

157 *Cf.* J. Bidez, *Eos ou Platon et l'Orient*, 1945, ch. 8

158 Both Sophocles and Euripides equate Hestia with the earth – Soph. fr. 615
(Pearson, vol. ii); Eur. fr. 938 (Nauck); *cf.* Tim. Locr. 97d; Plut., *De Primo
Frigido* 954f.

159 The extant work (itself a paraphrase of Plato's *Timaeus*) that goes under
his name is a first-century AD forgery

160 *Cf.* the two articles by G. Vlastos in *Studies in Plato's Metaphysics*, ed.
R. E. Allen, 1965, ch. 18, where most of the pertinent literature is cited

161 For which see Cornford's admirable commentary, pp. 59–72

162 In this famous phrase, it is a mistake to stress μῦθον at the expense of
εἰκότα – the notion of 'likeliness' or 'plausibility' is emphasized in 29c
(*cf.* Cornford, pp. 30ff.). It must be remembered that in Platonic doctrine
it is axiomatic that there can be no *true* knowledge of sensible phenomena;
anyone can make up an implausible account of the genesis of the universe
(*cf.* Hesiod's *Theogony* and the creation myths of the Egyptians and the
Babylonians), but it is the business of the philosopher to give the most
likely account consistent with his insight into the nature of real knowledge
(*cf. Tim.* 53d)

163 He even goes so far as to assert (p. 74), 'Plato probably had it before him
as he wrote'!

164 Properly speaking, an armillary sphere is a representation, by means of
circular rings mounted on a stand, of the main circles of the celestial
sphere (equator, tropics, zodiac, horizon, solstitial and equinoctial colures)
with the earth globe at the centre; it does not show the planetary orbits
or the fixed stars, except in so far as the constellations of the zodiac may
be marked on the band that represents it. An astrolabe is essentially a

sighting instrument used to determine the position of the sun or moon and the risings and settings of the more prominent fixed stars relative to a particular horizon (cf. my GFH, pp. 195–99). An orrery is a more complicated device that imitates the actual movements of the heavenly bodies round the earth by mechanical means (cf. R. T. Gunther, Early Science in Oxford, vol. ii, 1923, pp. 267f.)

165 Gunther, op cit., pp. 264ff. with illustrations

166 For ancient astronomical instruments in general, see my paper in Journ. Brit. Astron. Assoc., 64, 1954, pp. 77–85

167 Cf. Heath, Arist., pp. 162–63

168 Cf. Cornford (p. 103) quoting Aristotle, Phys. iv, 223b; for some criticisms of Cornford's remarks, see Vlastos in Studies in Plato's Metaphysics, pp. 409f.

169 Burnet's text reads εἰς [τὸν] τάχει μὲν ἰσόδρομον ἡλίῳ κύκλον ἰόντας, τὴν δ᾽ἐναντίαν εἰληχότας αὐτῷ δύναμιν. τόν is the unanimous reading of all the best MSS as well as Proclus and Stobaeus (and is retained by Hermann in the Teubner text and Rivaud in the Budé text), but its retention involves the attribution to Plato of the notion that the sun, Mercury, and Venus all move on one and the same circle, which is undoubtedly wrong in the present context and for Platonic astronomy in general; hence Burnet brackets it. Cornford (p. 105 note 2) suggests that τόν may have been read on the mistaken supposition that Plato believed that Mercury and Venus revolved round the sun (the theory attributed to Heracleides of Pontus – see p. 219), in which case all three bodies would have one main orbit. There is, however, another reading, τούς, which is given in the 'vulgate' and by a later hand in Y, and printed by Stallbaum and Bury (the latter in the Loeb Timaeus), and which gives the same sense as the omission of τόν – Taylor, (Tim., p. 196), though rejecting τούς, rightly construes ᾽εἰς (κύκλους) ἰόντας ἰσόδρομον ἡλίῳ κύκλον, κύκλον being an accusative of the internal object after ἰόντας

170 Cf. Heath, Arist., pp. 166ff.; Heath himself leaves the matter open

171 But see my explanation of this phrase, p. 112

172 Contrast this with the statement on pp. 86–7, 'We can now see why the changes in the relative positions of the planets are not ascribed merely to differences of speed, though that would be a possible way of representing the facts' (my italics)

173 Cornford rightly emphasizes (pp. 92–3; 109) that Plato is not writing a treatise on astronomy, but a myth of creation

174 This was generally accepted by the Greek astronomers; Ptolemy takes it for granted, and makes the mean, daily longitudinal movement of Venus and Mercury exactly the same as that of the mean sun (Almag. ix, 3 and 4). In Eudoxus' system also the same assumption is made, and Plato may well have derived his knowledge of the phenomenon from this source

175 Both Cicero (Tim. 25) and Chalcidius (Tim. 36d, p. 28 ed. Waszink) translate thus

176 And could readily be paralleled by similar instances in the comments of ancient and modern scholars unversed in scientific language – e.g. (*si parva licet* . . .) Guthrie (vol. i, p. 94), in discussing Anaximander's alleged astronomical system, actually speaks of a 'spherical plane'!!

177 This is the explanation favoured by Proclus (*in Tim.* 221e–f) after rejecting various others

178 Taken with laborious seriousness by Adam (vol. ii, App. I, pp. 264–312); but *cf.* Shorey's remarks (Loeb *Rep.*, vol. ii, Introd., p. xliv)

179 It is difficult to imagine what is meant by *all eight* orbits – i.e. including apparently the revolution of the Same, in terms of which all the others are measured, 39d6 – completing their courses together

180 *In Tim.* 276d *ad fin.* – followed by Heath (*Arist.*, p. 174) and Cornford (p. 119)

181 γῆν δὲ τροφὸν μὲν ἡμετέραν, εἰλλομένην δὲ περὶ τὸν διὰ παντὸς πόλον τεταμένον, φύλακα καὶ δημιουργὸν νυκτός τε καὶ ἡμέρας ἐμηχανήσατο, πρώτην καὶ πρεσβυτάτην θεῶν ὅσοι ἐντὸς οὐρανοῦ γεγόνασιν. This seems to be the soundest text. Burnet (*OCT*) reads . . . ἰλλομένην δὲ τὴν περὶ τὸν . . . and notes in his apparatus criticus '*b*8 ἰλλομένην FPr. Aristoteles Plut.: εἰλλομένην A: εἰλομένην P ci τὴν AP: om. FYPlut.' K. Burdach, however, in an article that deserves to be better known ('Die Lehre des Platonischen Timaios (40b) von der kosmischen Stellung der Erde', *Neue Jahrb. f. d. klass. Altertum* 49, 1922, pp. 254–78), after a careful and exhaustive survey of the various instances, shows that the form with ει is to be preferred here, and that the two word-families in εἰλ- and ἰλ- which originally had separate meanings (the former indicating 'press, crowd together' and the latter 'roll, wind, turn') became confused in Hellenistic times not only in spelling (both ει for ῑ and its converse being common in Alexandrine recensions), but also in meaning, so that in this case little weight can be attached to readings transmitted through literary sources; in particular, he shows that the notion that ἴλλω is an older and more genuine form of εἴλλω is entirely wrong. The retention of τήν in Burnet's text is indefensible, as Cornford points out (p. 120 note); it is not mentioned by any of the ancient commentators who quote the passage and is found in two MSS only

182 In both passages, the words translated 'coils and moves' are the same, ἴλλεσθαι καὶ κινεῖσθαι (which is the consensus of all modern editors), but the MSS also transmit the forms εἰλεῖσθαι, εἰλεῖσθαι, εἴλλεσθαι. Burnet's choice of ἰλλομένης at *Tim.* 40b–c is evidently influenced by these two passages in the *De Caelo*

183 *Cf.* H. Cherniss, *Aristotle's Criticism of Plato and the Academy*, 1944, Appendix VIII, pp. 557–58 – the whole of this Appendix is a valuable discussion of Aristotle's criticism with special reference to astronomical ideas

184 This is apparently Heath's final conclusion, *op. cit.*, p. 178

185 Cornford confuses his own argument by asserting, 'The effect is that in relation to absolute space she [the earth] stands still, while in relation to the other makers of day and night, the fixed stars, she rotates once every twenty-four hours in the reverse sense' (p. 131) – cf. p. 121, where he remarks more logically that 'the Earth must stand still, relatively to the diurnal revolution of the stars'

186 Cornford's citation (p. 130 note 3) of *Epinomis* 983b–c as evidence for a moving earth is misconceived because, as Cherniss points out (*op cit.*, pp. 556–57) γῆν τε καὶ οὐρανόν here 'is merely a solemn expression for "the whole material universe", the specific subject being the stars (including sun and moon) which carry out their revolutions in years, months, and days'

187 The weakness of which is demonstrated by the unconvincing nature of the scanty evidence adduced in support of it – e.g. when Timaeus Locrus describes the earth as ἐν μέσῳ ἱδρυμένα (97d), Cornford is constrained to remark 'a word which does not exclude motion' (p. 121 note 2) – Proclus knew better (*in Tim.* 281e). It is also difficult to reconcile a rotating earth with Cornford's own simile of the moving staircase (see above), which would seem to require the assumption of an absolutely stationary earth; it is noteworthy that he makes no attempt to explain this

188 Or, apparently, to Cornford – but Duhem (*Système du monde*, vol. ii, 1913, p. 88) read the sentence in this way

189 Cf. T. H. Martin (*Études sur le Timée*, vol. ii, pp. 88f.) who as long ago as 1841 gave what is essentially the correct interpretation

190 Cf. Proclus, *in Tim.* 281b *ad fin.*, τῷ οὐρανῷ δύναμιν ἔχουσά πως ἀντίρροπον, and, for night and day, 282c; Plut., *Quaest. Plat.* viii, 3, 1006e–f

191 *In Tim.* 281e *ad fin.*: 'Let Heraclides of Pontus . . . hold this opinion [viz. that the earth rotates], since he moves the earth in a circle; but Plato keeps it unmoved'

192 *In De Caelo*, p. 519, 9–11; cf. 444, 34ff.; 541, 28ff.

193 40d – whether we read ἄνευ διόψεως τούτων αὖ τῶν μιμημάτων . . . with most of the mss, or ἄνευ δι' ὄψεως . . . with Burnet (*OCT*) and Rivaud (Budé text), the sense is clearly that mere description of the phenomena without 'imitations' of the movements is insufficient for a proper understanding of them. However, as we have seen (pp. 120–21), such 'imitations' are not to be thought of as complicated models like the planetaria or orreries envisaged by Cornford, or the astrolabe mentioned by Proclus (*in Tim.* 248b); rather what is meant is a simple celestial globe that by its rotation can demonstrate risings and settings, or diagrams of the planetary movements such as Eudoxus must have used in constructing his system – μιμήματα can include drawing (cf. *Epin.* 975d)

194 40c, χορείας . . . καὶ παραβολὰς ἀλλήλων – Proclus (*in Tim.* 248c) correctly explains παραβολάς as 'comings together' in longitude, i.e. as applied to stars which rise or set simultaneously

195 The word is ἐπανακυκλήσεις. If my interpretation (above, p. 112) is correct, this means literally 'additional circlings', backwards in this case, i.e. retrograde movements of the planets in contrast to προχωρήσεις, 'progressions', forward motion along the zodiac; this is in accord with Proclus, who read ἀνακυκλήσεις and προσχωρήσεις (284a–c), but paraphrased by ὑποποδισμούς, 'retrogradations', and προποδισμούς, 'advances' (cf. Hypotyp. vii, 4). Cornford (p. 135), who notes that Heath's 'returnings of their orbits upon themselves' is unsatisfactory and himself translates 'counter-revolutions', nonetheless agrees that retrograde motion is meant (cf. above, pp. 132f.)

196 Apparently, even at this date, the identity of the Morning and Evening Stars with the single planet Venus was not generally recognized among non-scientists, despite the fact that, according to the doxographers, Parmenides had already pointed this out (see p. 51). The mention of the sun as also 'wandering' may perhaps refer to Eudoxus' (erroneous) belief that it exhibited deviations from the ecliptic in latitude (on this, see below); but, as so often in Plato, the text here is doubtful (cf. England's note ad loc.) and the meaning may be 'the sun and the moon do what we all know they do'

197 Cf. Heath's critical discussion of such views (Arist. pp. 182ff.)

198 As Taylor holds (Plato: The Laws, trans. A. E. Taylor, 1960, p. 210 note)

199 As suggested by Taylor (Plato: The Man and his Work, p. 486 note)

200 This is not to say that Plato knew the complete Eudoxan system, for which there is no evidence in Platonic astronomy; 'hints' of such knowledge found by Ross (Aristotle's Physics, 1936, p. 95) and Guthrie (Loeb De Caelo, pp. xix–xx, footnote) are chimerical – cf. Lasserre, Die Fragmente des Eudoxos von Knidos, 1966, pp. 181–82

201 On this concept, see Guthrie, Hist. of Greek Philos., vol. ii, pp. 163–64; cf. 414ff.

202 E.g. by Nicomachus (c. AD 100), Arith. i, 3, 5, ed. Hoche, Teubner, 1866

203 The single exception of any note (Diogenes Laertius' remark (iii, 1, 37) that 'some' attribute the Epinomis to Philippus of Opus is hardly good evidence for denying its Platonic authorship) is Proclus, who complains that it is 'full of spurious, mystical matter and betrays a foolish and senile mind' (in Remp. ii, p. 134, 5f. ed. Kroll), and who, according to the Prolegomena in Platonis Philosophiam ascribed to Olympiodorus (Hermann, Platonis Dialogi, Teubner, vol. vi, p. 218), put forward two very unconvincing arguments (one of which hinges on an astronomical point – see below) purporting to show that Plato could not have written the dialogue – on these arguments see E. des Places (Epinomis, Budé ed., 1956, p. 102)

204 For which see J. Harward, The Epinomis of Plato, 1928, pp. 26–58, and the edition of Des Places cited above, note 203, where most of the relevant literature is mentioned; cf. the latest editor F. Novotny, Platonis Epinomis, Prague, 1960. E. Dönt (see above, p. 235 note 141 – cf. also

Wiener Studien 78, 1965, p. 54) accepts it unquestioningly as a work written by Philippus of Opus – a position that few modern scholars would defend

205 *Gnomon* 25, 1953, pp. 371–75

206 Reviewing Novotny's ed. in *AJPh* 83, 1962, pp. 313–17

207 *Plato: the Man and his Work*, p. 498 note 1

208 *Cf. Timaeus* 47a–b for the same thought. At 47a5 Burnet (*OCT*, followed by Cornford) reads after ἐνιαυτῶν περίοδοι the words καὶ ἰσημερίαι καὶ τροπαί from one MS (F) although they are omitted by the two other best MSS (A and Y), and consequently by Hermann (and before him Stallbaum) in the Teubner text. Cicero (*Tim.*, ed. F. Pini, 1965, § 52) translates the Greek by 'annorumque conversiones' and evidently did not have equinoxes and solstices in his text; Chalcidius (§ 47, p. 44 ed. Waszink) has 'annorumque obitus et anfractus', which also suggests that he read ἐνιαυτῶν περίοδοι alone, without the addition of F's words (*cf.* his commentary, *op. cit.*, p. 156). In fact, the words καὶ ἰσημερίαι καὶ τροπαί, which add nothing essential to the sense, would seem to be an otiose gloss incorporated into the text of F

209 The sentiments here are very similar to those expressed in *Laws* 966–67 (see p. 141)

210 We should very much like to know what these 'proofs' were. Guesses about the size of the sun go back as far as Heraclitus, who thought it was only a foot wide (see p. 48), whereas Anaxagoras stated it was bigger than the Peloponnese (p. 58), and Archelaus that it was the largest of the celestial bodies (p. 77); but these are evidently no more than guesses. Later astronomers, as we shall see, used measurements of the earth's shadow at lunar eclipses to estimate the relative sizes and distances of sun, moon, and earth, and such methods may well have been known to Eudoxus who, according to Archimedes (*Arenarius*, p. 7 ed. Dijksterhuis) proved that the sun's diameter is nine times that of the moon. In all probability, then, it is to Eudoxus' 'proofs' that Plato refers

211 Much of this has already been said in *Laws* x – see above

212 The position of aether (αἰθήρ) here has been held to constitute an argument against the genuineness of the *Epinomis* (Cherniss in *Gnomon* 25, 1953, p. 372), on the ground that in this dialogue 'the ether is situated between fire and air', and 'since the regular solid which the *Epinomis* assigns to ether could still be only the dodecahedron and since the faces of this figure cannot be constructed out of Plato's two elementary triangles, the location of such a fifth body between fire and air would prevent the mutual interchange and transmutation of their corpuscles and so would disrupt the rationale of the "stereometric atomism" of the *Timaeus*'. The reference here is to *Tim.* 53c–57c where Plato gives his views as to how the basic elements combine to form different substances (see above, p. 232 note 122); according to Xenocrates (one of his pupils), Plato in the *Epinomis* intended

the dodecahedron to be the figure representing the fifth element, aether (Xenocrates fr. 53, Heinze). However, Cherniss' argument is invalidated by the simple fact that the aether, *as an element*, is *not* situated between fire and air in the *Epinomis*. It is true that at 984b6 we find αἰθέρα μὲν γὰρ μετὰ τὸ πῦρ θῶμεν, but this does not refer to the order of the elements, but to the order of the living beings whose main constituents are the elements; this is made absolutely clear by the preceding lines, νῦν οὖν δὴ περὶ θεῶν ἐγχειρῶμεν . . . τὰ δύο κατιδόντες ζῷα ὁρατὰ ἡμῖν, ἅ φαμεν τὸ μὲν ἀθάνατον, τὸ δὲ γήινον ἅπαν θνητὸν γεγονέναι, τὰ τρία τὰ μέσα τῶν πέντε τὰ μεταξὺ τούτων . . . πειραθῆναι λέγειν. The divine beings whose main constituent is aether are the first of the three classes intervening between the star gods and terrestrial life, but there is no reason to suppose that Plato intended this to be the actual order of the elements in the sense that Cherniss intends, nor that Plato or any of his readers would have found this statement inconsistent with the doctrine of the *Timaeus;* rather, it would seem logical to assign to two invisible classes of being, intermediate between the visible types, the appropriate invisible elements, aether and air, the former of which has already been postulated as the purest type of air in the *Phaedo* (109b; 111b) and *Timaeus* (58d). In any case, μετά at 984b6 may very well indicate not a spatial relationship, but a temporal one in the genesis of the universe, or an order of rank or importance; *cf.* 984c7, δεύτερα δὲ καὶ τρίτα καὶ τέταρτα καὶ πέμπτα ἀπὸ θεῶν τῶν φανερῶν ἀρξάμενα γενέσεως εἰς ἡμᾶς τοὺς ἀνθρώπους ἀποτελεὔτᾶν, and 984d. It does not necessarily follow that, because Plato chose to make aether the primary constituent of the second class in his hierarchy of living beings, this must reflect his conception of the physical order of the elements. Still less is it safe to assume (nor does Cherniss assume) that there is any spatial significance in the verbal order in which they are mentioned in the later dialogues; in the *Epinomis* itself (981c) they appear in the order fire, water, air, earth, aether; in the *Laws* (889b and 891c) the order is fire, water, earth, air; in the *Timaeus* the apparent order (to judge from 53c and 55d) is fire, earth, water, air – but in 53e we have one clear reference to spatial order when we are told that earth and fire are the extremes with the others (not actually named here) in between. Moreover, Cherniss' objection rests on the assumption that the dodecahedron is the regular solid to be assigned to the element aether; but there is nothing in the *Epinomis* to warrant this assumption, and Cherniss himself rightly remarks that this is probably Xenocrates' own 'attempt to read Aristotelian doctrine back into the *Timaeus*' (*loc. cit.*, note 4). The train of thought in the *Epinomis* is simply different from that in the *Timaeus;* in the former Plato is not concerned with the elements and their configuration out of 'atomic' triangles, nor with the interaction and transformation of elemental substances, but with a hierarchy of living organisms and their relationships to each other and to mankind – he is, in fact, operating on a different plane of imagination, and to demand a rigid correspondence in every detail between two such planes is to try to force a mechanical consistency on an intellect which has as many facets as there are stars in the heavens. The addition of aether as a fifth element in the *Epinomis* is certainly a novel development in Plato's thought (perhaps not entirely unheralded in the *Timaeus* – *cf.* Cornford, pp. 220–21, on 55c–d), but there is no justification for treating it as an argument against the genuineness of the dialogue

213 For a comparison of the demonology outlined here with that found in other Platonic dialogues (especially the *Symposium* and the *Phaedrus*) see the comments of Harward, Des Places, and Novotny *ad loc.*

214 On the other hand, there is plenty of imaginative fantasy to be found in the other Platonic dialogues, and in some ways (e.g. the importance of religious beliefs and of astronomy as an educational discipline) the *Epinomis* merely carries to a logical extreme tendencies and opinions already apparent in his earlier work. The senility of which Proclus complains (see above) may be more apparent than real, and although Plato was an old man when he wrote the dialogue, it would be presumptuous to decide that the new directions in which his thoughts were turning were unworthy of Platonic philosophy as a whole

215 A. Bouché-Leclercq, *L'astrologie grecque*, Paris, 1899, pp. 24, note 1 and *ff.*; 28*f.*; 75*f.*

216 *Cf.* M. P. Nilsson in *Harv. Theol. Rev.* 33, 1940, pp. 1–8, who points out that there were no indigenous cults of the heavenly bodies in early Greek history (Helios and Selene being merely gods of mythology), and that Plato was the first to insist on the divinity of the visible stars and suggest that they should be worshipped with full rites

217 *Cf.* my *GFH.*, pp. 12*f.* (where in footnote 4, for 'Kramer' read 'Cramer')

218 *Cf.* J. Bidez, *Eos ou Platon et l'Orient*, 1945. Bidez does not accept the Platonic authorship of the *Epinomis*

219 For the different types of astrology, see my article 'Astrology and astronomy in Horace', *Hermes* 91, 1963, pp. 67*f.*

220 The best MSS give (986b) μία δὲ τῶν πλανητῶν ἄστρων which is impossible since the sense demands that the reference here must be to the fixed stars, the planets being designated as πέντε δὲ ἕτεραι – hence Burnet (*OCT*) brackets πλανητῶν and Hermann (Teubner) reads ἀπλανῶν, but it is difficult to see why this easy reading should have been changed to πλανητῶν. Des Places (Budé) prints πάντων which has some MS support, and Novotny (*ad loc.*) is inclined to accept this reading

221 On planetary names see especially F. Cumont, 'Les noms des planètes et l'astrolatrie chez les Grecs', *L'Antiq. Class.* 4, 1935, pp. 5–43

222 *Cf.* Neugebauer, *Ex. Sci.*, pp. 80*ff.* It seems that the Egyptians could recognize some thirty-six constellations and stars (only two of which, Sirius and Orion, can be certainly identified with those familiar to us), later known as 'decans' (so called because they were supposed to rise above

the horizon at intervals of 10 days), which were used as a crude type of astronomical calendar; and they were interested in the heliacal rising of the bright star Sirius because it happened to coincide for some centuries with the annual flooding of the Nile, the main event in the life of the country. Apart from these few observational data there are no traces of any real astronomical concepts or of any underlying mathematical theory in old Egyptian texts. The Egyptian calendar, however, consisting of 12 months of 30 days each plus 5 additional days at the end of the year, became the standard astronomical time-scale used by the Greek astronomers (and still by Copernicus in the sixteenth century), and the Egyptian division of day and of night into twelve parts each also became standard practice – see below

223 This refers to the annual motion of the planets among the stars which takes place from west to east (opposite to the direction of the daily rotation of the heavens), and is therefore towards the right if one contemplates a conventional drawing or model of the celestial sphere with the north pole at the top (cf. Fig. 4, p. 18). Proclus (see above, note 203) tries to use this passage as an argument against the genuineness of the *Epinomis* by pointing to the apparent contradiction to the statement in the *Timaeus* that the revolution of the Same (i.e. the diurnal movement) is towards the right; but his argument is misconceived (in that he fails to appreciate the different viewpoints of the two passages – see pp. 121f.), and anyway this statement the *Epinomis* is consistent with *Laws* 760d where motion to the right is stated to be eastwards (τὸ δ' ἐπὶ δεξιὰ γιγνέσθω τὸ πρὸς ἔω)

224 Burnet (*OCT*), followed by Harward, actually inserts οὐκ, for which there is no MS authority at all, before ἄγων. Des Places, followed by Novotny, points out that this is unnecessary since the same sense can be obtained without the insertion of οὐκ – he translates 'pourrait avoir l'air d'entraîner les autres, du moins aux yeux des gens mal informés de ces questions'

225 This explanation was first suggested by Heath (*Arist.*, p. 185) and approved by Cornford (pp. 91–2). In his later (and slighter) book (*Greek Astronomy*, 1932, pp. xliii and 61–2) Heath seems to have accepted the other interpretation

226 The order given here is the order of the synodic periods, and also probably reflects the type of data found in Babylonian sources; the moon was the most important object of study for the Babylonian astronomers because their calendar was at all periods a strictly lunar one, the solstices were also tabulated, and tables are extant dealing with observations of Venus from the second millennium BC. If Plato had been speaking as a Greek professional astronomer, he would surely have placed the sun as the first of the seven orbits to be studied, in view of its fundamental importance as providing the basic time unit (the tropical year) and the basic line of reference (the ecliptic) in astronomical calculations

246 EARLY GREEK ASTRONOMY TO ARISTOTLE

227 A saying which Aristotle (*De Anima* 411*a*18) attributes to Thales – probably wrongly, *cf. CQ* 9, 1959, pp. 296–97

228 The geographical fragments have been edited by F. Gisinger, *Die Erdbeschreibung des Eudoxos von Knidos* (Stoicheia, Heft 6), 1921. The only collection of all the fragments is the recent edition (with commentary) by F. Lasserre, *Die Fragmente des Eudoxos von Knidos*, De Gruyter, Berlin, 1966; Lasserre's arrangement of the material leaves much to be desired as regards clarity, and his commentary is unilluminating for astronomical detail, but at least we now have all the relevant sources for Eudoxus (and much that is dubiously relevant) between the covers of one book

229 *Op. cit.*, pp. 5–6

230 G. de Santillana, 'Eudoxus and Plato', *Isis* 32, 1949, pp. 248–62; *cf.* Lasserre, pp. 137–42

231 In 365 according to De Santillana, or 360 according to F. W. F. von Bissing in *Forschungen und Fortschritte* 25, 1949, p. 225*f.*

232 E.g. Diog. Laert. viii, 87; Plutarch, *Marcell.* xiv, 11

233 Which we have on the excellent authority of Aristotle himself, *Nic. Eth.* i, 12, 1101*b*27*f.* and x, 2, 1172*b*9. In the latter passage Aristotle remarks on the temperance of Eudoxus' own character (διαφερόντως γὰρ ἐδόκει σώφρων εἶναι), which was not at all pleasure-loving, in spite of his philosophical opinions

234 E. Frank, 'Die Begründung der math. Naturwissenschaft durch Eudoxos' in *Wissen, Wollen, Glauben*, ed. L. Edelstein, 1955, pp. 134–57, especially 145–50

235 Apparently based solely on the undisputed fact that Eudoxus did at some period visit Sicily – *cf.* Aelian, *Var Hist.* vii, 17; Ptolemy, *Phaseis*, p. 67 ed. Heiberg, where he is stated to have made 'weather observations' (ἐπισημασίαι – see above, p. 85) in Asia, Sicily, and Italy

236 He was responsible for the redrafting of the theory of proportion, as set out and used in Euclid Books v and vi, to make it applicable to all magnitudes whether commensurable or incommensurable, and also for propounding the so-called 'method of exhaustion' for determining the areas and volumes of various curvilinear figures by 'exhausting' or using them up (δαπανᾶν) by inscribing polygons the areas of which were known – *cf.* Heath, *History of Greek Maths.* vol. i, pp. 322*ff.*; O. Becker, 'Eudoxos-Studien I-V' in *Quellen und Studien zur Geschichte der Math., Astron. und Physik*, Abt. B, Bd. 2, 1933, pp. 311–33 and 369–87; Bd. 3, 1936, pp. 236–44; 370–88; 389–410. Becker's attempt to construct a pre-Eudoxan general theory of proportion is criticized by Heath, *Maths. in Aristotle*, 1949, pp. 81–3

237 *In Arati et Eudoxi Phainomena commentariorum libri tres*, ed. Manitius, Teubner, 1894 – hereafter cited as *Comm. in Arat.*

238 This is more accurate than saying that the poem is simply a versification of Eudoxus, as I did on p. 1 of *GFH*

239 It needed proof because apparently many authorities doubted it, διὰ τὸ διστάζεσθαι παρὰ τοῖς πολλοῖς, i, 2, 1

240 For the calculation of latitude from the longest day, see Chapter I, pp. 19ff. The obliquity of the ecliptic in Eudoxus' time was 23°44' (see my GFH, p. 168), and this is the value I have used in the present calculations

241 Lasserre argues confidently for this (pp. 181ff.), to the extent of assigning fragments to the two works even when Hipparchus does not specify them by title (e.g. fr. 52 and 53) – on the assumption that the *Enoptron* contained better observations and 'stylistic improvements' (p. 192, *comment. ad loc.*)

242 5:3 gives 15 hours and 12:7 gives 15 hours 9 minutes; differences in the lengths of the day at the summer solstice were the means of differentiating latitudes north of the equator commonly used by pre-Hipparchian geographers – *cf. GFH*, pp. 159; 163

243 I.e. the great circles intersecting at the poles and passing through the solstitial and equinoctial points respectively of the ecliptic – called κόλουροι, i.e. 'curtailed', because their lower segments are cut off from view by the horizon (*cf.* Geminus, *Isag.* 5, 49)

244 Using such phrases as συμφώνως τοῖς φαινομένοις or συνεγγίζει τῷ φαινομένῳ (τῇ ἀληθείᾳ)

245 One might perhaps in these cases suspect textual corruption in the MSS of Eudoxus' works that Hipparchus used

246 In fact, Eudoxus was more correct here than Hipparchus; according to Baehr's tables, the declination of Canopus (α Carinae) in the former's time was −52.8° and in the latter's −52.7°, so that it would be near the limit of visibility at latitude 37° and invisible at the true latitude of Athens, 38°, but in both cases it would (just) be visible at Rhodes (*cf.* Geminus, *Isag.* 3, 15 and Manitius' note *ad loc.*)

247 These are probably κλ Draconis and 5 Urs. Min., the latter being a fourth magnitude star listed by Ptolemy among the ἀμόρφωτοι of Ursa Minor, i.e. those not counted in the actual constellation figure (*Synt.* ii, 38, 12 ed. Heib.)

248 On the role these played in ancient astronomy, see Chapter I, p. 19

249 *Cf.* ii, 1, 1 and especially ii, 1, 26, καὶ ὁ Εὔδοξος δέ, ᾧ κατακολούθηκεν ὁ ″Αρατος, τὸν αὐτὸν τρόπον ὑποτίθεται ἐν ταῖς συνανατολαῖς τὰς ἀρχὰς τῶν ζῳδίων ἐπὶ τῆς ἀνατολῆς – this evidently refers to a separate work of Eudoxus entitled *Simultaneous Risings* (Συνανατολαί) also apparently used by Aratus (*cf.* i, 4, 19 and i, 5, 15). Curiously, Lasserre ignores the evidence for such a work, although he is ready to attribute to Eudoxus another treatise with the improbable title *On Solar Occultations* (Περὶ ἀφανισμῶν ἡλιακῶν – *LSJ s.v.* ἀφανισμός also give this attribution) on very slender evidence (pp. 212–13)

250 See *JHS* 86, 1966, pp. 27–8 for the development of this usage

251 Stars not assigned letters by Bayer are usually designated by their numbers in Flamsteed's British Catalogue published in 1725

252 Cicero (*De Rep.* i, 22) quotes Gallus (a Roman astronomical writer of the second century BC) on Eudoxus' globe, *eandem illam sphaeram solidam astris quae caelo inhaererent esse descriptam*, and expressly contrasts this early type of solid globe with the *sphaera Archimedis*, which imitated the movements of sun, moon, and planets, and therefore was much more than a simple globe. Gallus is no great authority, it is true, since in the same passage he attributes the making of the first globe to Thales. . . .

253 Lasserre's contention (p. 191) that Eudoxus did not use a globe because the terms 'right' and 'left' are inappropriate for its surface is ill-conceived; one can perfectly well speak of 'the right' and 'the left' of *figures* on a globe (as the Greeks did), regardless of whether the directions are the same for the observer. Thus it makes no difference whether the figures are drawn as seen from the inside (as on celestial globes with the observer supposed to be at the centre), or as they actually appear in the sky relative to the right and left hand of the observer (as on many star maps); the left arm or the right foot or the head of the figure will in both cases denote the same stars. This was well understood by the Greek astronomers (*cf. Comm. in Arat.* i, 4, 9–11, where Hipparchus commends Aratus for attempting to give precision to his descriptions of figures by using the terms 'left' and 'right'); it is only modern commentators who have introduced confusion here. Where there was any likelihood of ambiguity, stars were described as lying further west, east, south, or north as the case might be

254 E. J. Webb, *The Names of the Stars*, 1952 (published posthumously) – see especially ch. 3–5 and *cf.* his remark on p. 63, 'If we look at the stars as they appear in the sky [a thing that very few scholars do, as Webb rightly complains] . . . we shall perceive in many cases . . . obvious reasons for names which have been quite obscured by the artificial figures, constructed often long ages after the names themselves had become traditional'

255 *Cf.* the information collected by M. P. Nilsson, *Primitive Time-Reckoning*, Lund, 1920

256 Theoretically, the vacant space should correspond to the 'antarctic' circle, i.e. the limit of the stars never visible at that time and place, its centre should mark the position of the south pole (because, owing to the effect of precession – see above, pp. 15*f.* this pole has shifted its position among the stars), and its radius the latitude of the constellation-makers

257 E. W. Maunder, *The Astronomy of the Bible*, 2nd ed., 1908, pp. 157–59

258 R. H. Allen, *Star Names and their Meanings*, 1899, pp. 14–15

259 Ptolemy says specifically that he himself has not always used the same shapes for the constellations as his predecessors, just as they did not always use the same as the astronomers before them, but made alterations in the interests of a more convenient arrangement, and he gives an example from Hipparchus (*Synt.* vii, 3, ed. Heib. vol. ii, p. 37, 11*ff.*). Modern astronomers from Bayer onwards have made further changes, until in 1930 by international agreement the present constellation boundaries were standardized

260 M. W. Ovenden, 'The Origin of the Constellations', *Philosophical Journal* 3(1), 1966, pp. 1–18. Ovenden seeks to demonstrate that the constellations were designed by the Minoans mainly as navigational aids round about 2,800 BC ± 300 years at latitude 36°N. ± 1½° and longitude 26⅓°E. – he has even found a suitable island in the Dodecanese for their observatory, the island of Stampalia (locally known as Astropalia 'which has an obvious astronomical "ring" about it')! Some of the arguments he uses are remarkably circular; in trying to prove (pp. 5–6) that certain constellations were arranged symmetrically with respect to the celestial north pole of that epoch, he uses a statistical method based on the hypothesis that they *were* so arranged as 'a band of the sky equidistant from the celestial pole' (and what? Ovenden does not say, but presumably means the equator). It is to be hoped that the 'results' of this fascinating paper will not be taken seriously; apart from the fallacious argumentation and fanciful speculation it contains, its whole thesis is vitiated by the totally unfounded assumption of the advanced astronomical knowledge possessed by the constellation-makers so-called – see below

261 *JHS* 86, 1966, p. 29

262 For the difficulties inherent in the concept of equinoxes, as opposed to solstices which are easily observable phenomena requiring no astronomical theory for their perception, see my article in *JHS* quoted above

263 Often made with the help of an instrument known as a 'precession globe', i.e. a celestial globe the poles of which are adjustable in circles round the ecliptic poles to take into account the shift in position (about 1° in 72 years) of the celestial poles owing to the effect of precession – see above. Even with this aid, such comparisons are of very doubtful validity, since we know neither the exact boundaries of the ancient constellations, nor the latitude of the original observations, nor the standard of accuracy involved; the latter especially, one suspects, is commonly overestimated by modern commentators

264 Robert Brown, *Researches into the Origin of the Primitive Constellations of the Greeks, Phoenicians and Babylonians*, 2 vols., 1899, 1900, p. 15 – a highly misleading work, packed with erroneous and outdated material. Only marginally less misleading is W. Hartner's article, 'The Earliest History of the Constellations in the Near East and the Motif of the Lion-Bull Combat', *JNES* 24, 1965, pp. 1–16, which, based on totally inadmissible premises, attributes sophisticated astronomical concepts to the Sumerians of the fourth millennium! Equally misguided are the attempts made to impute complicated astronomical motives to the builders of ancient monuments such as Stonehenge (e.g. by G. S. Hawkins in *Vistas in Astronomy*, vol. x, 1968, pp. 45–88); such fantasies are reminiscent of the 'pyramid literature' (on which see Neugebauer, *Ex. Sci.*, p. 96) – and equally valueless, despite the modern trappings of computer calculations with which they are invested

265 R. Böker, 'Die Entstehung der Sternsphare Arats', *Berichte über die Verhandlungen der sächsischen Akademie der Wissenschaften zu Leipzig* 99, 1952, pp. 3–68. Böker finds fault with Hipparchus' criticism of Eudoxus-Aratus because it is based on the supposition that the data were valid for Greece and that the colures were drawn through the beginning of the signs (p. 5); he completely ignores Hipparchus' own discussion of the latitude appropriate to the observations (*Comm. in Arat.* i, 3, 5–12), and his emphasis on the different placing of the solstitial and equinoctial points by Eudoxus and Aratus (ii, 1, 15*ff.*; 20*ff.*; 2, 5–6), which shows that he was fully alive to the difficulties of assessing the older material fairly. Böker's lack of historical sense is demonstrated by his reference to Aratus' astronomical source as 'Pseudo-Eudoxos' (duly castigated by Ludwig in *RE, s.v.* Aratos, Suppl. Bd. 10, 1965), his belief that the Greek constellations go back no further than the sixth century BC (thus apparently ignoring those mentioned in Homer and Hesiod – see above), his supposition that Eudoxus used the 360° division of the circle (in flat contradiction to the available evidence – see above), and his reference (p. 105 of a 'Nachtrag' to a German translation of Aratus by A. Schott – *Das Worte der Antike* VI, Munchen, 1958 – where Böker repeats much of the nonsense in his earlier paper) to Anaximander's famous sixth-century BC workshop in Miletus where he busied himself with all possible astronomical and meteorological instruments (!!). Moreover, Böker's supposedly scientific treatment of the Aratean data is basically unsound since he does not take into consideration the conditions and limitations of the original observational material (*cf.* Ludwig, *loc. cit.*), and anyway a close examination shows that out of the twelve Aratean passages he discusses in pp. 19–29, only three agree fully with the results obtained by his methods. Unfortunately, the erroneous conclusions he arrives at (that the Aratean sphere is valid only for the epoch – 1,000 ± 30–40 years at a latitude between 32°30′ and 33°40′, with the colures marked at the end of 15° of the relevant signs, and with the position of the zero point of the zodiac at about 26° of the ecliptic of AD 1900 – p. 8) have been accepted by Van der Waerden in his latest work, *Die Anfänge der Astronomie* (= Erwachende Wissenschaft II), Noordhoff, Groningen, 1966 – on which see further below

266 In the period between Hipparchus and Eudoxus the north celestial pole would have shifted westwards some 3°, thus altering the positions of the stars relative to the circles of the celestial sphere

267 On a very rough count, there are twenty-three such instances in Book i of the *Commentary* – admittedly, the disagreements are more than twice as numerous, but then its main purpose is to criticize and correct

268 This passage alone is enough to refute a large part of Ovenden's thesis, since he makes much play with alleged differences in the respective 'zones of avoidance' of the Eudoxan-Aratean and the Hipparchian spheres (*op. cit.*, pp. 9–10)

269 *Cf.* Neugebauer, *Ex. Sci.*, pp. 103*ff.*

270 'Babylonian' is used as a convenient generic term for the distinctive culture of the Tigris and Euphrates valleys, which was dominated at different times by Akkadians, Kassites, Assyrians, Persians, and finally Macedonians and Greeks in the Seleucid period

271 *Cf.* the omen series of texts collectively known as 'Enuma Anu Enlil' discussed by E. F. Weidner in *Archiv für Orientforschung* 14, 1942 and 17, 1954

272 *Cf.* Van der Waerden, *Anf.*, pp. 56*ff.*

273 All authorities agree that these 'ways' are bands of a certain width (variously estimated) parallel to the equator, and not lines delimiting zones as in the Greek concept of 'arctic' and 'antarctic' circles – *cf.* C. Bezold, A. Kopff and F. Boll, 'Zenit- und Aequatorialgestirne am babylonischen Fixsternhimmel', *Sitzber. d. Heidelb. Akad. d. Wiss.*, phil.-hist. Kl., Abh. 11, 1913, pp. 3–59; E. F. Weidner, 'Ein babylonisches Kompendium der Himmelskunde', *Amer. Journ. Sem. Lang. and Lit.* 40, 1924, pp. 186–208, and 'Der Tierkreis und die Wege am Himmel', *Arch. f. Orientf.* 7(4), 1931, pp. 170*ff.*; J. Schaumberger, 3. *Ergänzungshefte zur Sternkunde und Sterndienst in Babel*, Kugler, 1935, pp. 321*ff.*

274 Rome, 1950 = Teil 4, Bd. 2 of the *Sumerisches Lexikon*, ed. P. A. Deimel

275 Particularly striking is the fact that the Babylonians designated the 'horn' (i.e. the claws) of the Scorpion as ZI.BA.AN.NA/zibanîtu, meaning 'Balance, Scales' (Gössmann, p. 72), just as the Greek astronomers differentiated the Claws(Χηλαί, Latin *Chelae*) from the rest of the constellation and later called them the Balance (Ζυγός, Latin *Libra*) – the latter name seems to be post-Hipparchian, for in the *Comm. in Arat.* Χηλαί is consistently used except in one passage (iii, 1, 5), which Manitius regards as spurious on other grounds as well (p. 303, note 41). Ptolemy (*Synt.* viii, 1, *ad init.*) uses, Χηλαί for the *figure*, but Ζυγός for the *sign* – *cf.* Bouché-Leclercq, p. 141

276 Van der Waerden, *Anf.*, pp. 67–8

277 *Op. cit.*, pp. 256–57; but Van der Waerden himself admits that this does not hold for Aries, Cancer, and Aquarius, and his arguments for Virgo (*cf.* Webb, *The Names of the Stars*, p. 33, 'the assumption that, because the Greek Virgin carries a Corn-Ear in her hand, the Babylonian Corn-Ear must have been carried in the hand of a Virgin, though apparently taken for granted by all Assyriologists, is of course ridiculous'), Sagittarius, and Capricornus are extremely flimsy, depending largely in the last two cases on representations of the figures in the Dendera zodiac, which since it dates from the Roman period in Egypt, is hardly convincing evidence

278 This is against the view supported by Webb (see above, pp. 159–60) that certain names (including the Triangle) are obviously appropriate for certain star-groups; Van der Waerden (p. 68) specifically comments on the likeness of these particular stars to a plough!

279 *Cf.* Van der Waerden, p. 68 and diagram p. 66

280 The Seleucid era began in 312 BC; after this there is a long series of texts
(the latest being dated to AD 75) which show Babylonian mathematical
astronomy at its highest level of development – see O. Neugebauer,
Astronomical Cuneiform Texts, 3 vols., 1955. There is no doubt at all, as we
shall see, that the results obtained and (to some extent) the methods used
by the Babylonians during this period were known to the Greek
astronomers from at least the time of Hipparchus (second century BC)
onwards. The disputed questions are when and how these results and
methods were transmitted to Greece, and what influence the earlier stages
of Babylonian astronomy had on its Greek counterpart, and again when
and how this influence was exerted

281 Aristophanes fr. 163, πόλος τόδ' ἐστίν; εἶτα πόστην ἥλιος τέτραπται;
Even here it is not altogether impossible for πόλος to have the same
meaning as in *Birds* 179f., namely 'region of the sky'

282 In which case it would be similar to the 'scaphe' (σκάφη) mentioned by
Cleomedes (*Cycl. Theor.* i, 10, 54f.) as being used by Eratosthenes for his
famous measurement of the earth. Vitruvius in his chapter on sun-dials
(ix, 8) speaks of 'a hemisphere hollowed out of a square' (*hemicyclium
excavatum ex quadrato*) which was supposed to have been invented by
Berosus (on whom see note 306, below); but this must have been a later
type because he describes it as 'cut away to suit the latitude' (*ad enclimaque
succisum*), and this presupposes greater theoretical knowledge than either
Berosus or Eratosthenes could have possessed – see *CQ* n.s. 5, 1955, pp.
248f. Vitruvius goes on to say that the invention of the 'scaphe or hemisphere' was attributed to Aristarchus

283 For common misconceptions concerning the use of sun-dials in antiquity,
see *JHS* 86, 1966, p. 29

284 *Cf.* Diels, *Antike Technik*, 3rd ed., 1924, pp. 162–63

285 *Cf.* Neugebauer, *Proc. Amer. Philos. Soc.* 107, 1963, p. 533

286 *Cf.* Weidner, *Amer. Journ. Sem. Lang. and Lit.* 40, 1924, pp. 198f.; Neugebauer, *Isis* 37, 1947, pp. 37–43; Van der Waerden, *Anf.*, pp. 80–1

287 Neugebauer, *Ex. Sci.*, pp. 81; 85–6. Van der Waerden claims (*op. cit.*, p. 88)
that the division of day and night into twelve equal periods each is attested
by the numbers on an ivory prism of the Assyrian period (thus before
630 BC); but his interpretation of this text (following Fotheringham in
The Observatory, No. 703, 1932, p. 338) is far from secure, and on his own
admission the meaning of half of it remains unknown

288 To explain the frequent Greek references to Egyptian astronomical
observations, Van der Waerden suggests that in the period from 630 to
480 BC (during part of which Egypt came under Babylonian rule)
Babylonian astronomical ideas strongly influenced the local Egyptian
astronomy, to the extent of causing a resurgence of observational activity,
and that this was how Eudoxus came to profit from his stay in Egypt
(*Anf.*, pp. 130f.). The evidence for such a supposition is extremely thin,

and consists (apart from inferences drawn from Greek literary sources) of a single Demotic text, written in the first century AD but apparently based on an original of the late-sixth or early-fifth century BC, which contains a number of eclipse omens arranged by the months in which they take place, groups of three months being assigned to four separate terrestrial regions, and also a concordance of Egyptian and Babylonian names for the months. However, since the contents of the text obviously belong to the pre-scientific stage of astronomy and there is no mention of the planets or the fixed stars, Van der Waerden is forced to assume a later blooming of Egyptian observational astronomy (*op. cit.*, p. 133), for which, needless to say, there is no evidence at all

289 See Bouché-Leclercq, *L'astrol. grecque*, p. 93, note 2
290 In the Philolaic system and perhaps by Democritus – see pp. 65f.; 82
291 See the Venus tablets of Ammizaduga in the astrological omen series 'Enuma Anu Enlil' – Van der Waerden, *Anf.*, pp. 34ff.
292 Published in *Late Babylonian Astronomical and Related Texts*, 1955, (Brown University Studies 18), by A. J. Sachs as facsimiles of the original copies made by Pinches and Strassmaier – no translations are given
293 Van der Waerden, *Anf.*, p. 105
294 See the standard work by O. Neugebauer, *Astronomical Cuneiform Texts*, 3 vols. 1955, where all the Seleucid material is dealt with
295 *Cf.* Van der Waerden, *Anf.*, p. 166
296 *ACT*, vol. ii, p. 280
297 *Cf.* Neugebauer, *Proc. Amer. Philos. Soc.* 98, 1954, p. 64: 'But there is no trace of any definition of the vernal point as the intersection of ecliptic and equator (which nowhere appears in Babylonian astronomy)'
298 *Cf.* Van der Waerden, *Anf.*, pp. 104–05
299 *ACT*, vol. ii, p. 281
300 Neugebauer, *Journ. Cuneif. Stud.* 2, 1948, pp. 209ff.
301 Neugebauer, *Journ. Amer. Orient. Soc.* 70, 1950, pp. 1–8; *cf.* Van der Waerden, *Anf.*, pp. 115–16
302 Van der Waerden, *Arch. f. Orientf.* 16, 1953, p. 223
303 A. Sachs, *Journ. Cuneif. Stud.* 2, 1948, pp. 289–90; Van der Waerden, *loc. cit.*, p. 222, note 25
304 Fotheringham remarks that this puts Ptolemy in a better position than any modern astronomer as regards the length of observation series available to him (*The Observatory*, No. 51, 1928, pp. 312–13)
305 On this see especially Neugebauer, *Proc. Amer. Philos. Soc.* 107, 1963, pp. 534–35
306 The Babylonian priest Berosus (or Berossos), who founded a school in the island of Cos and wrote in Greek a history of his own country (see the fragments edited by P. Schnabel, *Berossos und die bab.-hell. Lit.*, Leipzig, 1923), is often cast in the role of intermediary between Babylonian and Greek science (*cf.* Neugebauer, *Ex. Sci.*, p. 157). Unfortunately, the

extant fragments do not bear this out, and what little astronomy they contain (e.g. fr. 16-26 on the phases of the moon) bears no relation to contemporary Babylonian lunar theory (Neugebauer, *Proc. Amer. Philos. Soc.* 107, 1963, p. 529). Berosus dedicated his *Babyloniaca* to Antiochus I (281-261 BC) and is thus, anyway, too late for Eudoxus

307 *Proc. Amer. Philos. Soc.* 107, 1963, pp. 529*ff.*

308 Neugebauer, *Ex. Sci.*, pp. 102; 140. Van der Waerden thinks differently – see below

309 *Cf. ACT*, vol. i, p. 11, 'All that can be said with safety at present is that the methods for computing lunar and planetary ephemerides were in existence *c.* 250 BC. Their previous history is unknown to me'

310 *Journ. Cuneif. Stud.* 6, 1952, pp. 54*ff.* It should be noted that Sachs' treatment depends on the *assumption* 'that the planetary data refer to signs of the zodiac, not constellations' (p. 55)

311 Even Van der Waerden admits that there is no trace of the zodiac as such in these texts – *Anf.*, p. 77 (see also below)

312 Chiefly in his book *Die Anfänge der Astronomie* (= Erwachende Wissenschaft II), Noordhoff, Groningen, 1966, which sums up the results of his earlier papers on the subject

313 *Op. cit.*, pp. 171-72; 201-03

314 See *JHS* 86, 1966, p. 29

315 On which see above, p. 43 and *CQ* 9, 1959, pp. 294*ff.*

316 See his highly misleading account of Pythagorean astronomy in *Die Astronomie der Pythagoreer*, Amsterdam, 1951

317 To the examples mentioned above, add his unquestioning acceptance of Böker's untenable theory that the Eudoxan-Aratean sphere is only accurate for a date about -1,000 BC (see above, p. 162 and note 265)

318 According to the *Ars Eudoxi* (see above, p. 88, note 109), Eudoxus estimated the number of days from the autumnal equinox to the winter solstice as 92 and from the winter solstice to the vernal equinox as 91; but unfortunately the papyrus is defective with regard to his estimates of the other two astronomical seasons (cols. 22-3). Since it is certain that in the Eudoxan solar theory the sun's longitudinal motion is assumed to be uniform (see above, p. 181), Heath is probably right in supposing that Eudoxus made the lengths of the seasons 91 days each, with an additional day for autumn to make the total up to 365 (Heath, *Arist.*, p. 200), thus making no use of the discovery by Meton and Euctemon of the inequality of the seasons (see below, p. 88)

319 E. F. Weidner, *Arch. f. Orientf.* 7(4), 1931, p. 171; O. Neugebauer, *Isis* 37, 1947, p. 38

320 Kugler, *Sternkunde u. Sterndienst in Babel*, vol. i, 1907, p. 13; vol. ii, 1909-10, pp. 77-8; *cf.* Neugebauer, *Ex. Sci.*, p. 169

321 *Cf.* T. W. Africa, *Science and the State in Greece and Rome*, 1968, p. 37: 'Hamstrung by the dogma that celestial motion was perfect and circular,

Greek astronomers expended great ingenuity to reconcile the erratic
behavior of the planets with their presumed circular motion'

322 Proclus in his *Comment. in Euclid. i* expressly draws attention to the fact
that several of the theorems were included because of their usefulness in
astronomy (pp. 268–69 ed. Friedlein). The earliest extant Greek mathematical treatise, Autolycus' *On the Moving Sphere*, dated to the last decades of
the fourth century BC, already contains propositions relating to the sphere
which are merely stated without proof, and were therefore presumably
taken from a still earlier textbook on 'sphaeric' which contained the proofs –
cf. Heath, *Hist. of Greek Maths.*, vol. i, pp. 349–50

323 Neugebauer, *Ex. Sci.*, p. 110

324 Delambre, for example, in his still indispensable *Histoire de l'astronomie
ancienne*, 2 vols., 1817, nowhere mentions the planetary scheme, and only
deals with Eudoxus' other astronomical work in the course of a chapter on
Aratus (tom. i, ch. 4, pp. 61–74), although he has a brief section (tom. i,
pp. 301–10) on Simplicius' *in De Caelo*. Delambre evidently did not know
of Eudoxus' mathematical work, since he says (p. 131), '. . . rien ne
prouve qu'il fût géomètre'

325 L. Ideler, *Abh. d. Berlin. Akad.*, hist.-phil. Kl., 1828, pp. 189–212, and
1830, pp. 49–88; E. F. Apelt, *Abh. d. Fries'schen Schule*, Heft 2, Leipzig,
1849; T. H. Martin, *Mém. de l'Acad. des Inscript. et Belles-Lettres*, tom. 30,
pt. 1, 1881, pp. 153–302; G. V. Schiaparelli, 'Le sfere omocentriche di
Eudosso, di Callippo e di Aristotele', *Pubblic. del R. Osserv. di Brera in
Milano 9*, 1875 – German trans. by W. Horn, *Abh. z. Gesch. d. Math.*,
Heft 1, Leipzig, 1877. pp. 101–98. Martin states (pp. 160–61) that his own
work was completed before Schiaparelli's, and that the reading of the
latter's description has not caused him to make any changes in his own
views, which, as we shall see, differ from Schiaparelli's in one important
respect

326 J. L. E. Dreyer, *A History of Astronomy from Thales to Kepler*, Dover repr.,
1953, pp. 89–103; T. L. Heath, *Aristarchus of Samos*, pp. 193–211

327 In describing the rest of Eudoxus' system, it will henceforth be taken for
granted that each sphere is affected by the rotations of the spheres enveloping
it; for economy of words only the individual rotations will be mentioned

328 Simplic., p. 495,4 ed. Heib., ἐγκεκλιμένος πρὸς τὸν διὰ μέσων τῶν ζῳδίων
τοσοῦτον, ὅσον ἡ πλείστη κατὰ πλάτος τῇ σελήνῃ παραχώρησις γίγνεται
. . . τὴν τρίτην δὲ [ὑπέθετο] διὰ τὸ μὴ ἐν τοῖς αὐτοῖς τοῦ ζῳδιακοῦ
σημείοις βορειοτάτην τε καὶ νοτιωτάτην φαίνεσθαι γινομένην, ἀλλὰ
μεταπίπτειν τὰ τοιαῦτα σημεῖα τῶν ζῳδίων ἀεὶ ἐπὶ τὰ προηγούμενα

329 τὴν μετάπτωσιν παντάπασιν ὀλίγην γίνεσθαι καθ' ἕκαστον μῆνα, p. 495,
14

330 In fact, its changes in *declination* (from the equator) are much greater than
its changes in *latitude*, but Eudoxus probably failed to distinguish between
the two – *cf.* Martin, *op. cit.*, p. 217

331 Martin (pp. 214–15) suggests the sidereal or tropical month of about 27½ days, but there is no evidence and little likelihood that this was known to Eudoxus

332 Whatever speed was assigned to the third sphere (and conjecture is fruitless) could easily have been compensated by increasing slightly the speed of the second sphere

333 E.g. W. D. Ross, *Aristotle's Metaphysics*, vol. ii, 1924, pp. 385*ff.*

334 The nodes are the two diametrically opposite points where the mean lunar orbit intersects the ecliptic

335 Who clearly indicates in *Met.* Λ 8, 1073*b*26–7 that the second sphere for all the planets and the sun and moon represents their direct (eastwards) motion along the ecliptic

336 Namely, the slow rotation of the poles of the lunar orbit round the poles of the ecliptic. The only evidence adduced for Eudoxus' knowledge of this period is a remark by Ptolemy that the eclipse period of 223 lunations was known 'roughly' to 'the still older astronomers' (ὁλοσχερέστερον μὲν οὖν οἱ ἔτι παλαιότεροι . . . ἔγγιστα ἑώρων μῆνας μὲν ἀποτελουμένους σκγ, *Almag.* iv, 2, p, 270, 1*ff.* ed. Heib.), and since he counts Hipparchus as παλαιός (iii, 1, p. 191, 17*f.*), παλαιότερος is taken as referring to Eudoxus. This is very tenuous, and anyway the context makes it clear that οἱ ἔτι παλαιότεροι here refers to the Babylonian astronomers (οἱ Χαλδαϊκοί – *op. cit.*, p. 270, 20), whose results Hipparchus discussed

337 *Cf.* Delambre, tom. i, pp. 73; 122; 125

338 Martin realized this clearly (*op. cit.*, especially pp. 216–21) and his treatment of the Eudoxan system is sensible apart from a tendency to accept data derived from the *Ars Eudoxi* as genuinely Eudoxan when there is no proof of this; Lasserre also notes the uncritical assumption of a mistake on the part of Simplicius (*Die Fragmente des Eudoxos von Knidos*, 1966, p. 202), but gives no discussion of the issues involved

339 This is so on both Schiaparelli's and Simplicius' interpretations

340 Εὐδόξῳ τοίνυν καὶ τοῖς πρὸ αὐτοῦ τρεῖς ὁ ἥλιος ἐδόκει κινεῖσθαι κινήσεις, p. 493, 11*f.*

341 Theon Smyrn., pp. 135 ed. Hiller (apparently from Adrastus – *cf.* p. 129); 173; 194; Pliny, *Nat. Hist.* ii, 67; Alex. Aphrod., *in Met.* 8, p. 703 ed. Hayduck; Chalcid., *Comm. in Tim.* 77, p. 125 ed. Waszink; Mart. Cap. viii, 849 (287 G), p. 315 ed. Eyssenhardt; also 863 (293 G), p. 322 ed. Eyssenhardt

342 The mathematical details are best studied in Schiaparelli or Heath (who also gives a modern solution, using analytic geometry, by N. Herz); *cf.* also Neugebauer, *Scripta Math.* 19, 1953, pp. 226–29

343 Since the figure is described on the surface of a sphere, Schiaparelli calls it (somewhat misleadingly) a 'spherical lemniscate'

344 *Comm. in Eucl.* i, ed. Friedlein, Teubner, pp. 127, 1; 128, 5

345 The ancient astronomers regularly assumed that the sidereal periods of

Mercury and Venus were the same as the sun's, i.e. 1 year (see note 174), because these planets are never seen far from it. In fact, this assumption results from the very large parallax effects caused by the earth's rotation in the case of the inferior planets, whereas for the superior planets (at their much greater distances from the sun) such effects are far less noticeable. On the geocentric hypothesis one is bound to give Mercury and Venus sidereal periods equal to that of the sun, and no valid comparison with the true values (on the heliocentric system) can be made

346 The *Ars Eudoxi* actually gives the figure of 116 days for the synodic period of Mercury (col. 5, p. 16 ed. Blass), but there is no knowing whether this was in fact derived from Eudoxus or a later source

347 It should be emphasized that, in this part of his reconstruction, Schiaparelli is demonstrating his own ingenuity rather than that of Eudoxus (as Martin points out – *op. cit.*, p. 225 note 1). There is no evidence that Eudoxus used the figures given by Schiaparelli and, to judge from the inaccuracy of much of the rest of his astronomical work, it would seem highly improbable that he knew the correct values

348 The logic of this last assertion is questionable, since in fact the maximum latitudes are approximately 2½° for Saturn and 1½° for Jupiter

349 If indeed he assumed any as regards the inclinations and the dimensions of the 'hippopedes'. It is by no means impossible that he contented himself with showing the theoretical possibility of explaining retrograde motion and stationary points by means of the 'hippopede', without actually assigning any parameters to the systems apart from the sidereal and synodic periods

350 *Cf.* Dreyer, pp. 38off.

351 It does not seem that Eudoxus regarded his sets of spheres as anything other than mathematical abstractions; there is no evidence that he speculated on the material of which they were comprised or the connection between them or the power that moved them. As we shall see, Aristotle was concerned with all these things. According to Archimedes (*Aren.* 9, ed. Heib., vol. ii, p. 220) Eudoxus supposed the sun's diameter to be nine times that of the moon; how he arrived at this figure we do not know – in reality the ratio is about 400:1

352 καὶ δοκεῖ μάλιστα πάντων αὕτη ἡ περίοδος τοῖς φαινομένοις συμφωνεῖν, *loc. cit. ad fin.*

353 On the 'parapegmata', see above, pp. 84f.

354 Latitudinal zones on the terrestrial sphere in which, for practical purposes, such data as the length of the longest day, the ratios of the gnomon to its shadow at stated times, and the appearance of the night sky remained the same for all observers in the same 'clima' – see on the development of this concept *CQ* 5, 1955, pp. 248ff.; *CQ* 6, 1956, pp. 243ff.; *GFH*, pp. 154ff.

355 Vitruvius (ix, 8) says in connection with sun-dials that Eudoxus (or, according to some sources, Apollonius) invented the 'spider' (*arachne*).

What this was is not certain, but it is not unlikely that Eudoxus investigated the shadows cast by a gnomon, and the lines and circles marking these may have suggested the term (*cf.* Diels, *Antike Technik*, 3rd ed., 1924, pp. 160–61 and diagram on p. 163). Later, the movable disc (representing the ecliptic) of the planispheric astrolabe was called the 'spider (*GFH*, pp. 197; 201), but Eudoxus certainly did not know this instrument

356 Ptolemy says (*Phas*, p 67, 5 ed. Heib.) that Callippus made observations in the Hellespont

357 Sosigenes (a Peripatetic philosopher of the second century AD – *not* the astronomer of the same name who helped Julius Caesar reform the calendar; Simplicius makes it clear that Sosigenes derived most of his information from Eudemus' *History of Astronomy*, on which *cf.* CQ 9, 1959, pp. 301*f*.) *ap.* Simplic. p. 504, 17, οὐ μὴν αἵ γε τῶν περὶ Εὔδοξον σῴζουσι τὰ φαινόμενα, οὐχ ὅπως τὰ ὕστερον καταληφθέντα, ἀλλ'οὐδὲ τὰ πρότερον γνωσθέντα καὶ ὑπ'αὐτῶν ἐκείνων πιστευθέντα

358 *Loc. cit* 36–7, τὰ φαινόμενα εἰ μέλλει τις ἀποδώσειν – a perfectly regular and straightforward use of ἀποδίδωμι in the sense of 'account for', 'explain' (see *LSJ s.v.*). Unfortunately, Sosigenes (followed by all later commentators) preferred the far less accurate, if more picturesque, phrase τὰ φαινόμενα σῴζειν, i.e. 'to preserve (agreement with) the facts of observation'. Kranz (*Rhein. Mus.* 100, 1957, p. 128) is wrong in supposing that the phrase was first used by Heracleides Ponticus, a pupil of Plato; the passage he quotes from Simplicius in support of this is clearly an *indirect* quotation and describes Heracleides' views in Simplicius' own words – *in De Caelo* ii, 13 p. 519 ed. Heib., ἐν τῷ κέντρῳ δὲ οὖσαν τὴν γῆν καὶ κύκλῳ κινουμένην, τὸν δὲ οὐρανὸν ἠρεμοῦντα Ἡρακλείδης ὁ Ποντικὸς ὑποθέμενος σῴζειν ᾤετο τὰ φαινόμενα. Sosigenes' phrase, in its literal English translation of 'to save (preserve) the appearances' with all the ambiguities inherent in the expression, has led to the misleading idea that the Greek astronomers were concerned mainly with distorting the results of observation to make them fit into preconceived, theoretical schemes. The whole history of Greek astronomy, which shows a steady development from the naiveties of the Pre-Socratics, through the Pythagoreans and Plato, to the system of Eudoxus, and finally to the Hipparchian-Ptolemaic system of epicycles and eccentrics, demonstrates how false this idea is; at each stage, as new and more accurate observations were accumulated, older theories were dropped in favour of newer ones which seemed to provide a more complete explanation of astronomical facts

359 p. 497, 17 ed. Heib.

360 Cols. 22–3; these values are a considerable improvement on those of Euctemon (*cf.* p. 88) and, compared with the true figures for 330 BC, are less than half a day out – *cf.* Schiaparelli, p. 46

361 The mathematical details are given by Schiaparelli (*op. cit.*), whom Heath (*Arist.*, pp. 213–16) follows closely. It must again be emphasized (*cf.* note

347) that the values assumed by Schiaparelli are entirely conjectural, as indeed is the mode of operation of the additional spheres, since we have no information on these points from the ancient sources

362 Geminus, *Isag.* 8, 58–60; *cf.* p. 189 above

363 As explicitly stated by Hipparchus *ap.* Ptol., *Almag.* iii, 1, p. 207, 11 ed. Heib.

364 E.g. *Almag.* iii, 1; iv, 10; v, 3; vi, 5; vii, 3, *et al.*

365 *Cf.* Van der Waerden, *JHS*, 80, 1960, p. 170; Ginzel, *RE*, Bd. 10(2), 1919, col. 1663

366 Ed. W. K. C. Guthrie, Loeb, repr. 1960; P. Moraux, Budé, 1965

367 *De Caelo* ii, 10, 291α29–32, περὶ δὲ τῆς τάξεως αὐτῶν [τῶν ἄστρων], ὃν μὲν τρόπον ἕκαστον κεῖται (*v.l.* κινεῖται) τῷ τὰ μὲν εἶναι πρότερα τὰ δ' ὕστερα, καὶ πῶς ἔχει πρὸς ἄλληλα τοῖς ἀποστήμασιν, ἐκ τῶν περὶ ἀστρολογίαν θεωρείσθω: *cf.* 291b10 and *Met.* Λ 8, 1073b3–6

368 W. D. Ross, *Aristotle's Physics*, 1936, Introd., pp. xv–xviii

369 W. D. Ross, *Aristotle's Metaphysics*, 1924, Introd., pp. xxiv–xxix

370 *Op. cit.*, pp. xxix note 1; 382; 384. It is fair to point out that the difference in styles has also been given the opposite interpretation, i.e. that the fuller style is *earlier* than the more concise one (e.g. by F. Blass in *Rhein. Mus.* 30, 1875, pp. 481–505). P. Merlan (*Traditio* 4, 1946, pp. 1–30), on the other hand, sees 1074a31–38 as an essential part of the argument of this chapter, which he finds 'logically and satisfactorily organized' (*op. cit.*, p. 14)

371 *Cf.* Moraux, Introd., pp. lxiv–lxv; cxxiv*ff.*

372 Thus various attempts have been made to trace specifically Aristotelian doctrines, such as those of the fifth element and the Unmoved Mover, back to the lost, early dialogue *De Philosophia* (of which we have fragments, ed. V. Rose, Teubner, pp. 24–40), and to discover a line of development between this (presumed to be still strongly under Platonic influence) and Aristotle's later ideas; but the very divergent conclusions reached by different scholars (for references, see Moraux, pp. li–lv) do not inspire much confidence

373 Except Aristarchus and Seleucus, who undoubtedly put forward at least tentatively a heliocentric hypothesis – see Heath, *Arist.*, pp. 301*ff.*

374 For the astronomical opinions of many of the Pre-Socratics, Aristotle's is the only evidence that can be considered to any degree reliable – see above, Chapter III

375 In the whole work, Aristotle records only one detailed astronomical observation (that of an occultation of Mars by the moon, ii, 12, 292a3–6, τὴν γὰρ σελήνην ἑωράκαμεν – hence presumably observed by Aristotle himself – διχότομον μὲν οὖσαν, ὑπελθοῦσαν δὲ τῶν ἀστέρων τὸν τοῦ Ἄρεος, κ.τ.λ.), and this is neither dated by him (modern calculations show that it was probably the occultation of 4 May 357 BC – *cf.* Guthrie *ad loc.*, Loeb, p. 205) nor reported in the form that a practising astronomer would use (compare the manner in which similar occultations by the moon of

the Pleiades are reported by Ptolemy from Timocharis, 283 BC, and Agrippa, AD 92, in *Almag.* vii, 3, ed. Heib., vol. ii, p. 25, 15*f.* and p. 27, 1*f.*). Aristotle adds that similar observations have been made by the Egyptians and the Babylonians (see above, p. 167)

376 *Cf. Met.* Λ 8, 1073*b*9–10, πλείους γὰρ ἕκαστον φέρεται μιᾶς τῶν πλανω-μένων ἄστρων. I follow Cherniss' interpretation (*Aristotle's Criticism of Plato and the Academy*, 1944, App. VIII, pp. 547*f.*) which is surely correct. Aristotle has at the back of his mind those like Heracleides Ponticus who were able to account for the daily risings and settings by postulating a rotating earth and a fixed outer sphere, but he insists that this is not sufficient, since any planetary body must have more than one *individual* motion; but, if the earth *is* given the requisite number of motions, then other phenomena will be produced which are contrary to the facts of observation. This seems preferable to assuming (with Heath, *Arist.*, p. 241, and Guthrie, *op. cit.*, pp. 242–43) that Aristotle was unaware of the alternative explanation of the daily revolution (a highly improbable assumption, since Heracleides was also a pupil of Plato), and was referring only to a double motion of the earth, one component of which he infers must be in the plane of the ecliptic

377 See above, pp. 136*f.*

378 On τροπαί, see p. 116

379 κυρτός, 297*b*28; *cf. Meteor.* ii, 7, 365*a*32, ὡς οὔσης [τῆς γῆς] κυρτῆς καὶ σφαιροειδοῦς – strictly, this proves only the curvature of the earth's surface, but taken in conjunction with the other arguments it serves as proof of the sphericity

380 The examples he gives, that stars seen in Egypt and Cyprus are invisible further north and that those continuously visible in the north are seen, further south, to rise and set (298*a*3–6), are almost certainly taken from Eudoxus – *cf.* p. 155

381 The figures cited by Guthrie (Loeb, pp. 254–55) from Prantl are completely and inexplicably wrong (except the last one), and it is incredible that they should continue to be repeated (e.g. by J. H. Randall, *Aristotle*, p. 160, note 21). For the conversion of stades into miles, the best assumption is to take 8.72 stades as equivalent to 1 English mile or (which gives roughly the same result) 10 stades as equivalent to 1 geographical or nautical mile (= 1.152 Eng. miles) – on the value of the stade, see *GFH*, pp. 42–6

382 *De Caelo* ii, 11, 291*b*13, ἡ δὲ φύσις οὐδὲν ἀλόγως οὐδὲ μάτην ποιεῖ: *cf.* ii, 8, 289*b*25; 290*a*31. This is the keynote of all Aristotle's natural philosophy – he is a teleologist

383 It is difficult to imagine what gave rise to this curious notion, which (as far as I know) does not appear in any other ancient source. Simplicius in his commentary (pp. 454–56 ed. Heib.) is obviously (and not surprisingly) unhappy about Aristotle's arguments in the whole of this section; he quotes the passage in the *Timaeus*, and refers approvingly to Ptolemy's opinion

that the stars *do* rotate since it is in their nature to do so (*Hypoth. Planet.* ii, p. 131, 9*f*. ed. Heib., vol. ii, *Claud. Ptol. Op. astron. min.* – this second book of the *Planetary Hypotheses* is extant only in an Arabic translation of which Heiberg gives a German version by L. Nix) – but he throws no further light on any actual observations of this alleged phenomenon

384 This is not stated explicitly, but seems a necessary inference from his previous words, since he is evidently using ἄστρον as a general term to denote *all* the celestial bodies – *cf.* 290a14–15, μόνος δὲ δοκεῖ τῶν ἄστρων ὁ ἥλιος . . . Heath (*Arist.*, p. 235) attempts a defence of Aristotle which carries little conviction

385 This is an essential presupposition for the theory of unmoved movers (see below), and Aristotle takes great pains in arguing against the proponents of a plurality of worlds, the Atomists

386 αἰθήρ, which he derives from ἀεὶ θεῖν, 'always running' (*cf.* Plato, *Crat.* 410b), because it is in continual motion, criticizing Anaxagoras' use of the term to denote fire (*De Caelo* i, 3, 270b22–5; *cf. Meteor.* i, 3, 339b22). Yet in *Phys.* iv, 4, 212b21 Aristotle himself uses it as a synonym for πῦρ. For αἰθήρ in Homer, see above p. 30

387 1074a7–8, τούτων δὲ μόνας οὐ δεῖ ἀνελιχθῆναι ἐν αἷς τὸ κατωτάτω τεταγμένον φέρεται. Actually, as Sosigenes notes (*ap.* Simpl., p. 503, 28), without discussing the point, the moon also requires counteracting spheres; for Aristotle's own explanation in the *Meteorologica* of such phenomena as comets, shooting stars, and the Milky Way (on which see below) envisages the outer layer of the sublunary sphere as being carried round in the same way as the fixed stars – so that there is the same need for the motions of the moon's individual spheres to be cancelled out as in the cases of the other planetary bodies (*cf.* Heath, *Arist.*, p. 219)

388 1074a12–14, εἰ δὲ τῇ σελήνῃ τε καὶ τῷ ἡλίῳ μὴ προστιθείη τις ἃς εἴπομεν κινήσεις, αἱ πᾶσαι σφαῖραι ἔσονται† ἑπτά τε καὶ τεσσαράκοντα

389 1074a3, εἰς τὸ αὐτὸ ἀποκαθιστάσας τῇ θέσει. Sosigenes spells this out – speaking of the last of Saturn's counteracting spheres, he says (*ap.* Simpl., p. 502, 17) στραφήσεται οὖν οὕτως ὁμοίως κινουμένη τῇ ἀπλανεῖ, οὐ μέντοι καὶ τὴν τάξιν ἕξει τῆς ἀπλανοῦς, περὶ ἄλλους στρεφομένη πόλους καὶ οὐ τοὺς τῆς ἀπλανοῦς – *cf.* p. 498, 5–7

390 The same error vitiates the results of an ingenious paper by N. R. Hanson, 'On Counting Aristotle's Spheres', *Scientia*, 98, 1963, pp. 223–32, who tries to prove that a 55-sphere system cannot work, but that 'The required observations can be generated either within a system of 49 spheres, or within one comprising 61 spheres – the latter being preferable from an "Aristotelian" point of view' (p. 223); later he suggests that either 50 or 54 or 62 or 66 spheres would suffice (p. 232)! Hanson agrees that '55 spheres is all right [*sic*] if only we assume that each new α [the first sphere in each set] *absolutely* has the motion of the fixed stars' (p. 229), but he regards this as incompatible with Aristotle's 'unified mechanically-articulated

cosmology'. Yet this assumption *does* according to all our evidence underlie Eudoxus' original scheme (*cf. Met.* Λ 8,1073*b*18–19, ὧν τὴν μὲν πρώτην τὴν τῶν ἀπλανῶν ἄστρων εἶναι) and Hanson rightly points out that 'Aristotle is explicit in adapting [*sic*] *en bloc* Eudoxus' technique for his own cosmology' (p. 226). His ideas about the connections postulated between the spheres by Aristotle go well beyond what we are entitled to infer from Aristotle's words, and the whole paper (which out-Aristotles Aristotle) seems an excellent example of the dangers of forcing a spurious scientific rigidity on the modes of ancient astronomical thought. However, Hanson is probably more correct than he realizes in saying (p. 229) that, if 55 spheres are insisted upon, 'Aristotle's entire cosmology becomes a childish re-scaling of the Eudoxan calculation technique' – although 'childish' seems unnecessarily harsh

391 287*a*5–11, especially the last sentence, ὥστε σφαιροειδὴς ἂν εἴη πᾶσα· πάντα γὰρ ἅπτεται καὶ συνεχῆ ἐστὶ ταῖς σφαίραις

392 οὐρανός in all three of the senses which Aristotle defines in *De Caelo* i, 9, 278*b*8*ff.* (namely, the outer circumference of the world, the celestial regions, and the whole universe) is conceived of as 'body' (σῶμα); *cf.* ii, 3, 286*a*10–12

393 ii, 12, 293*a*6–8, ἐν πολλαῖς γὰρ σφαίραις ἡ τελευταία σφαῖρα ἐνδεδεμένη φέρεται, ἑκάστη δὲ σφαῖρα σῶμά τι τυγχάνει ὄν

394 See *De Gen. et Corr.* ii,10, 336*a*31*ff.*, διὸ καὶ οὐχ ἡ πρώτη φορὰ αἰτία ἐστὶ γενέσεως καὶ φθορᾶς, ἀλλ' ἡ κατὰ τὸν λοξὸν κύκλον, and the whole of this chapter; also *Meteor.* i, 9, 346*b*21–4; ii, 4, 361*a*12–14; *cf. Met.* Λ 5, 1071*a*13, . . . ὥσπερ ἀνθρώπου αἴτιον τά τε στοιχεῖα . . . καὶ παρὰ ταῦτα ὁ ἥλιος καὶ ὁ λοξὸς κύκλος, where Ross' note (p. 365) to the effect that it was Hipparchus who first called the ecliptic ὁ ἐκλειπτικός is wrong – the latter term is not found until Achilles Tatius in the third century AD (its appearance in Cleomedes, *Cycl. Theor.* ii, 5, p. 206, 26 ed. Ziegler, is an interpolation), whereas Hipparchus and Ptolemy always use the phrases ὁ λοξὸς κύκλος or ὁ διὰ μέσων τῶν ζῳδίων κύκλος (as does Aristotle himself – see above and Λ 8, 1073*b*19), restricting ἐκλειπτικός to the meaning 'pertaining to eclipses'

395 ἐλάττους γὰρ ἥλιος καὶ σελήνη κινοῦνται κινήσεις ἢ τῶν πλανωμένων ἄστρων ἔνια, 291*b*35

396 *Phys.* vii, 2, 243*a*32–4, where the mover is said to be ἅμα the moved and ἅμα is defined as 'nothing being in between them' (ὅτι οὐδέν ἐστιν αὐτῶν μεταξύ) – a similar definition of ἅμα is given in v, 3, 226*b*21; *cf.* viii, 1, 242*b*59–63

397 258*b*10–11, ἀνάγκη εἶναί τι ἀΐδιον ὃ πρῶτον κινεῖ, εἴτε ἓν εἴτε πλείω. Later Aristotle says that it is better to envisage only one unmoved first mover, on the principle of the economy of hypotheses (259*a*6–13); but, as we shall see, he has to abandon this position with regard to the movers of the planetary spheres

398 As Ross explains (*Arist. Phys.*, pp. 727–28), 'the sphere may be regarded

as produced by uniform expansion from the centre, or by uniform contraction from the circumference'

399 Ross (*Arist. Met.*, p. cxxxiv) says, 'This, however, is an incautious expression which should not be pressed. Aristotle's genuine view undoubtedly is that the prime mover is not in space', and he cites in support *De Caelo* 279a18. Yet this passage does not altogether bear out Ross' opinion. Aristotle is explaining (279a6ff.) that there cannot be any bodily mass outside the heavens, for the world (ὁ πᾶς κόσμος) is made up of all the available matter (ὕλη) and there is only one world; 'outside the heaven, there is neither place nor void nor time (οὐδὲ τόπος οὐδὲ κένον οὐδὲ χρόνος ἐστὶν ἔξω τοῦ οὐρανοῦ) . . . and therefore neither are the things there born in place (διόπερ οὔτ' ἐν τόπῳ τἀκεῖ πέφυκεν), nor does time make them grow old, nor is there any change at all in any of the things that are posited to lie beyond the outermost revolution (τῶν ὑπὲρ τὴν ἐξωτάτω τεταγμένων φοράν), but changeless and unaffected they continue to lead the best and most self-sufficient life throughout all eternity' – and Aristotle goes on to stress the ideas of immortality and divinity which men have always connected with the notion of eternity. Thus there *are* at least conceptual entities ἔξω τοῦ οὐρανοῦ – and what better place could there be for the notional first mover or movers? *Cf.* W. Theiler in *JHS* 77 (1), 1957, pp. 127–31, who cites two passages in Sextus Empiricus (*Hyp.* iii, 218 and *Adv. Math.* x, 33) for Aristotle's view of god πέρας τοῦ οὐρανοῦ. There is no need at all to assume 'an incautious expression' on Aristotle's part. In fact, the eternity of the whole heaven (including all time and infinity) is described in this passage of the *De Caelo* in terms that recall those applied later to the prime mover (it is 'deathless and divine' and the source of existence and life for all other things, 279a28–30; *cf.* *Met.* Λ 7, 1072b14 and 28–30); the reason why the latter is not actually mentioned by name here is presumably because Aristotle had not yet elaborated this concept (*cf.* Cherniss, *Arist. Crit. Pl. and Acad.*, App. X, p. 588). P. Merlan in *Apeiron*, Monash University, Australia, 1, 1966, pp. 3–13, has an interesting analysis of this passage; his main thesis, that Aristotle's theology is basically polytheistic, is probably correct; less convincing is his insistence that Aristotle's views on the fifth element and the unmoved mover or movers are self-contradictory – on this apparent contradiction, see further above, p. 213

400 1071b20, ἔτι τοίνυν ταύτας δεῖ τὰς οὐσίας εἶναι ἄνευ ὕλης· ἀΐδίους γὰρ δεῖ, εἴπερ γε καὶ ἄλλο τι ἀΐδιον. ἐνέργειαι ἄρα – note the *plural*, which anticipates the prime movers of the planetary spheres mentioned later; hence the plural ἐνέργειαι (found in two MSS) seems better than the singular of the Oxford text (*cf.* Merlan, *loc. cit.*, p. 11 note 1)

401 κινεῖ δὲ ὡς ἐρώμενον, 1072b3. For both desire (ὄρεξις) and thinking (νόησις) are types of movement in Aristotle's view (*cf.* 1072a30, νοῦς δὲ ὑπὸ τοῦ νοητοῦ κινεῖται), as is explained in *De Anima* iii, 10

402 1072b26–30; cf. Nic. Eth. x, 8, 1178b21–2

403 It is difficult to see how to reconcile this statement with the ἀνελίττουσαι σφαῖραι

404 This is the very puzzling passage 1074a31–8 (see above, p. 195). One may agree with Merlan (*Traditio* 4, 1946, p. 13) that Aristotle is arguing against views such as those attributed to Heraclides Ponticus (Aët. ii, 13, 15; *Dox. Gr.*, p. 343), that every star was a world by itself, with its own earth and air; but why does he, immediately after stressing the plurality of prime movers (and since these are 55 or 47 in number, how is it that they do not partake of matter by Aristotle's own reasoning?), relapse into speaking of 'the first unmoved mover' (τὸ πρῶτον κινοῦν ἀκίνητον ὄν, a37) as if it were the only one? A possible explanation perhaps (which I have not seen suggested elsewhere) is that Aristotle regarded the prime mover of the outer heaven as a *primum inter paria;* there is only one basic principle of celestial movement (ἀρχή) for our one, unique universe, but this principle manifests itself in a number of independent prime movers of which the first (activating the sphere of the fixed stars) is commonly used as the exemplar, since the daily revolution of the heavens from east to west is the only revolution common to all the celestial bodies. Such an explanation (which cannot here be developed in detail) might serve to account for the emphasis that Aristotle lays (both here and in *De Caelo* i, 8 and 9 – see above, p. 199) on the idea of one world only; more than one world would entail more than one set of unmoved movers, and more than one principle of circular motion would be utterly incompatible with his whole philosophical system

405 1074a38–b14, esp. b2, θεοί τέ εἰσιν οὗτοι καὶ περιέχει τὸ θεῖον τὴν ὅλην φύσιν and b9, ὅτι θεοὺς ᾠοντο τὰς πρώτας οὐσίας εἶναι, θείως ἂν εἰρῆσθαι νομίσειεν. If we regard a31–38 as a long parenthesis (see above, p. 195), the antecedent of οὗτοι will be the divine celestial bodies mentioned in a30; but even if we do not so regard it, the lack of a specifically expressed antecedent need not prevent its referring to the prime movers which have been the subject of discussion (Merlan – *op. cit.*, p. 14 – thinks of a gesture towards the heavens when the passage was read aloud). It seems likely that by τῶν ἀρχαίων καὶ παμπαλαίων (b1) and τῶν πρώτων (b14) Aristotle means the ancient Egyptians and Babylonians (cf. *De Caelo* ii, 12, 292a7–9), who undoubtedly had an astral religion from very early times; but there is no evidence for star gods in early Greek belief

406 This and the last two books (iii and iv) of *De Caelo* deal with the four elements, their mutual transformations, and the general principles of the processes of generation and decay – cf. H. D. P. Lee, *Meteorologica*, Loeb, 1952, p. x

407 This occurrence was known to Anaxagoras, who is even supposed to have predicted it (*DK* 59 A11 and 12; Diog. Laert. ii, 10), but this is simply a picturesque inference derived from his well-known views of the celestial

bodies as fiery stones (see above, pp. 58f. and Heath, *Arist.*, p. 246 note 1)

408 This comet is mentioned three times by Aristotle (343b1; b18; 344b34), always with the epithet 'great' (ὁ μέγας). It is possible that it was an appearance of what is now known as Halley's comet, which in its 76- or 75-year period would have been due to appear about 370 BC

409 This is a common phenomenon in naked-eye observation, and its precise description by Aristotle confirms that he himself had experience of it. The star in question is either δ or ε Canis maioris

410 If this is what is referred to in *Meteor.* i, 5 as Ideler, Heath, and Lee assume; E. W. Webster (in the Oxford translation, vol. iii, 1931, *ad loc.*) thinks rather of 'phenomena of cloud coloration' – *cf.* Lee, p. 36. The aurora is rarely seen except in extreme northern or southern latitudes round the earth's north or south magnetic poles

411 And therefore rises and sets with the heavens – hence the moon also should have been endowed with counteracting spheres (see note 387)

412 See pp. 29f. for this concept in Homer

413 Aristotle's 'explanation' is even less convincing than usual. He seems to regard the Milky Way as that part of the sphere of the fixed stars which contains the greatest number of bright stars (346a17ff. – hence presumably the appellation 'greatest circle'); but the fixed stars form the outermost sphere of the whole universe, and it is difficult to see how this can be regarded as in close enough contact with the outer stratum of the sublunary region to ignite it. Perhaps he would invoke the concept of the counteracting planetary spheres (see above) to meet this point. It is interesting that the text refers to a diagram (ὑπογραφή) and a globe (σφαῖρα) on which stars might be marked (346a32f.) – no doubt as visual aids to accompany the lecture (*cf.* Lee, p. 67 note *b*)

414 See pp. 84ff.

415 ὁ διὰ παντὸς φανερὸς (κύκλος), 362b2 – but this, as Aristotle must have known (*cf. De Caelo* ii, 14, 297b31f.), properly refers to the limit of the circumpolar stars at a particular latitude, which changes with the observer's locality. Poseidonius and Strabo rightly criticize Aristotle for defining a zone by a variable circle (Str. C 95 – see *GFH*, p. 166). The same confusion is apparent in the chapter on winds (ii, 6), where in the circular diagram based on the eight commonly used reference points (namely, summer and winter sunset and sunrise, equinoctial sunset and sunrise, and the north and south poles – see Lee, p. 187), a chord connecting the points where two northerly winds blow is described as 'nearly corresponding to the wholly visible circle, but not accurately' (ἡ δὲ τοῦ ΙΚ διάμετρος βούλεται μὲν κατὰ τὸν διὰ παντὸς εἶναι φαινόμενον, οὐκ ἀκριβοῖ δέ)

416 This is because of his insistence that motion must start from the right, therefore east must be the right-hand side, and the motion of the heavens must be from right to left. However, when we face south (as in the northern hemisphere we must in order to face the sun), the motion of the heavens

is clearly from our left (east) to our right (west), on the normal supposition that we are standing up with our heads in the direction of the visible north pole; but this does not accord with the right's being the start of the motion – hence we have to suppose that our *feet* are towards the north pole and our head towards the south, which is therefore the upper pole. This whole chapter shows Aristotle at his least convincing in an astronomical context. Wicksteed (Loeb *Physics*, pp. lxii–lxiii) has a good note on this passage, and correctly points out that in terrestrial maps east is on the right, but in celestial maps (e.g. *Norton's Star Atlas*) east is on the left. For Pythagorean views on the supremacy of the right and Plato's connection of this with the east, see above p. 121

417 From what we are told of Aristotle's relations with Callippus, *Met.* Λ 8 can hardly have been written before 330 BC and probably later – see pp. 190 and 194

418 E.g. *Met.* A 9, 992a32; *cf.* his sharp criticism in *De Caelo* iii, 7 of Plato's elemental triangles as described in the *Timaeus* – see Solmsen, *Arist. Syst. Phys. World*, 1960, pp. 259f.

419 *Cf.* Merlan in *Traditio* 4, 1946, p. 5, who draws attention to the tripartite classification mentioned in *Phys.* ii, 7, 198a29, whereof astronomy belongs to the second part which is concerned with 'things that are in motion but are indestructible'

420 See p. 199; *cf. Met.* Λ 8, 1050b22, where Aristotle assures us that there is no need to fear that the heavens will become tired!

421 E.g. ii, 6, 288a27–b7; 288b22–30; iv, 3, 311a9–12; *cf.* i, 8, 277b9–12 and Guthrie *ad loc.*

422 The controversy has centred on *De Caelo* i, 9, 279a33–b3; ii, 1, 284a18–b4; ii, 3, 286a10–12; iii, 2, 300b18–22; iv, 2, 309b17–24 – all these passages have been thought to be inconsistent with the concept of a prime mover – as well as on the passages mentioned above. Cherniss (*loc. cit.*) cites most of the relevant literature; *cf.* Moraux, p. xliii–xlv

423 For the stars do not vary their distances from each other – 288b10–12; *cf.* 289a5–7

424 287a23–4, ἡ τοῦ οὐρανοῦ φορά . . . μόνη συνεχὴς καὶ ὁμαλὴς καὶ ἀΐδιος

425 ἀνωμαλία . . . ἀνωμαλίαν: μεταβάλλοι . . . μεταβάλλει; ἀδυναμία . . . ἀδύνατον (288a26; 288b5; 288b21)

426 As Cherniss points out, *op. cit.*, App. X, pp. 581–82; *cf.* Moraux, p. xlv

427 *Cf. Met.* Θ 8, 1050b21, οὐκ ἔστι κατὰ δύναμιν κινούμενον ἀλλ᾽ ἢ ποθὲν ποί

428 287b27–8, ἀνάγκη γὰρ καὶ τοῦτο ἢ ἀρχὴν εἶναι ἢ εἶναι αὐτοῦ ἀρχήν

429 The passages cited by Ross (*Met.* Λ 4, 1070b34; 7, 1072b13; 10, 1076a4) are simply general expressions of the universal influence of the prime mover of the outer heaven to which, in a sense, everything in the celestial regions is subject, since it is the cause of all risings and settings. As mentioned above (note 404), Aristotle seems to use this as an exemplar or typical prime mover

430 The question cannot be argued in detail here, but *cf.* the views of Merlan cited above, note 399

431 *Cf.* Moraux, p. xlii and lxxxviii and the authorities referred to in his notes

432 This itself is incompatible with the description in ii, 4 (see above, p. 203) – air has no business to be in the celestial regions at all, and the only planetary spheres immediately beneath which there is air are those of the moon (*cf.* Meteor. i, 3 and 4; above, pp. 199; 203)

433 This is part of his argument against the Pythagorean 'harmony of the spheres'

434 291b2, ἕκαστον γὰρ ἀντιφέρεται τῷ οὐρανῷ κατὰ τὸν αὑτοῦ κύκλον – this, of course, refers to motion along the zodiac in the opposite direction (i.e. west to east) to the daily rotation

435 Thus Saturn has the longest sidereal period (about 29½ years) and the moon the shortest (about 27⅓ days)

436 *Cf.* Moraux, p. cii, 'Les théories proprement astronomiques du *De Caelo* ne paraissent ni très claires ni très cohérentes'; *cf.* pp. cxxv-cxxvi

437 E.g. *De Caelo* ii, 3, 286a4–7; 5, 287b31–288a2; 12, 291b25–8; 292a14–18; *Met.* Λ 8, 1073b13–17

438 Hence, of course, the comparative success of the Greeks in the two fields, mathematics and astronomy, which make least use of the controlled experiment

439 And Ptolemy was just as convinced of the divine nature of the heavens and the celestial bodies as ever Plato and Aristotle were (*cf. Almag.* i, 1, p. 6, 23 ed. Heib., and the famous epigram in the *Palatine Anthology*, ix, 577)

440 *Summa Theol.* i, 50, 3–4; *cf.* P. Duhem, *Le système du monde*, tom. v, 1917, p. 539$ff.$

441 See P. Wicksteed, *The Reactions between Dogma and Philosophy*, 1920, pp. 33$f.$

442 In reality, we know nothing for certain of how Eudoxus or Callippus made their observations. It is a reasonable assumption that they made use of celestial globes and the gnomon, but we can only guess at the type of sighting instrument they employed, if indeed they used any – it is remarkable what can be achieved by simply using the fingers of the outstretched arm to gauge the relative positions of celestial objects

443 Author of two extant treatises, *On the Moving Sphere* and *On Risings and Settings*, which (with Euclid's *Phainomena*) are the earliest mathematically based works on astronomy that have come down to us – but there is no actual mention of the theory of concentric spheres in these works

INDEX

Adam, J., 107, 233
Aegospotami: fall of meterorite at, 78, 208
ἀήρ, 30
Agathemerus, 83
Air: as ἀρχή, 46, 77
αἰθήρ ('aether'), 30, 51, 53, 58, 142, 144, 199, 203, 211, 242f., 261
Alcmaeon, 74
Alcman, 225
Allen, R. H., 160, 162, 248
Anaxagoras, 52, 55–9, 77, 78, 82–3, 94–5, 99, 199, 205, 208, 227–8, 261
Anaximander, 42, 44–5, 48, 50, 90, 174, 226
Anaximenes, 42, 46–7, 77, 95
A'ntarctic' circle, 155, 163
Antiphon, 90
Apelt, E. F., 177, 255
Aquinas, St Thomas, 134, 217
Arabs, 218
Aratus, 153–4
ἀρχή, 44f., 64, 77, 205, 206, 207, 264
Archelaus, 77
Archimedes, 23
Archytas, 76f., 198
'Arctic' circle, 31, 49, 155
Arcturus, 36f., 89, 164
Arctus (Great Bear), 30–1, 49
Aries, 16, 17
Aristarchus, 23
Aristophanes, 89, 141, 165, 227, 252
Aristotle, 7, 40, 47, 53, 54, 55, 167, 193–218, 225,

227, 259–66; on Pythagoras, 63f.; on Pythagoreans, 64f., 67f., 72f., 230; on 'harmony of the spheres', 71f.; on Democritus, 82; on Plato, 101f., 132–6, 216f., 233, 239; on Eudoxus, 177; on Callippus, 190f.; as observer, 196, 208f., 259–60, 264; universe of, 199f., 203f., 261, 262; doctrine of natural motion, 196, 199, 211f., 213, 259f., 266; modification of system of planetary spheres, 200ff., 261; on moved and movers, 205f., 213f., 262, 263, 266; on position of poles, 210, 265–6; inconsistencies in astronomical views of, 211ff.
Ars Eudoxi, 88, 191, 231, 254
Ascension: right, 16
Astrolabe, 120, 237, 238
Astrology, 8, 27, 28, 29, 34, 145, 168–9, 221, 244
Astronomy: development of, 27f., 50, 60, 106f., 171f., 174f., 217, 221, 267; Egyptian, 28, 46, 161, 167, 244–5, 252–3; Babylonian, 14, 146, 161, 163–75, 225, 245, 251–4; Minoan and Mycenaean, 29, 161; in Platonic education, 92f., 99, 105f., 137f., 141, 148, 231, 245; Greek words for, 207; Aristotelian view of, 211f.

Athens: latitude of, 154, 209, 221
Atoms, 79; and Atomists, 140
Aurora, 209, 265
Autolycus, 218, 221

Babylonian astronomy, 14, 146, 161, 163–75, 225, 245, 251–4; units of measurement, 166
Baehr, U., 209, 224
Berosus, 253–4
Boeckh, A., 138–9
Böker, R., 162, 250
Bouché-Leclercq, A., 145, 244
Brown, R., 162, 249
Burdach, K., 132, 239

Caesar, Julius, 85
Calendar, 75, 93; in Geminus' Isagoge, 84, 230; establishment of, 29, 85f., 188f.; confused state of Athenian, 89; Egyptian, 166–7, 244–5; Babylonian, 85f., 169
Callippus, 85, 189, 190–3, 201–2
Cancer, 17, 210
Canopus, 155, 163, 247
Capricorn, 17
Cardinal points, 32
Cave: simile of, 99f., 106f., 235
Censorinus, 75, 87, 88, 188
Chalcidius, 120
Cherniss, H., 212, 242–3, 260

268

272 INDEX

Sphere: celestial, 17–18, 61, 88, 89, 113, 117, 149, 153, 161, 166, 169, 173, 177–9, 189–96; of universe, 51, 53, 177; of fixed stars, 66, 68–9, 83–4, 147; 'harmony of the -s', 71, 75; as shape of celestial bodies and earth, 72, 74, 196, 197, 203, 257; atomic components of, 82; in *Phaedo*, 96, 232; armillary, 120; of Archimedes, 120, 159; concentric, 177–8, 190, 203; counteracting, 200, 216

Spiral: of planetary motions, 125, 129, 132, 139

Stars: 'fixed', 10–11, 47, 55, 68–9, 83–4, 131–2, 146–7, 171, 215; in Homer, 30; nature of, 46–7, 48, 51, 55, 59, 78, 131; revolution of, 59, 83; size of, 143; description of, 158; rotation of, 131, 198; shooting, 209

στεφάναι, 50

Stewart, J. A., 97

Strabo, 49

στρογγύλος, 78, 83

Sun: annual motion of, 16–17, 18, 83–4, 88, 157; 'turnings' of, 17–18, 31, 59, 78, 94, 116, 130–1, 210; cult of, 28, 29, 221; size of, 48, 55, 58, 77, 143, 242; hemispherical, 55; composition of, 47, 48, 51, 59, 68, 78, 82, 90; fire of, 90; in Eudoxan system, 181–3; in Callippic system, 191; rotation of, 198

Sundial, 165, 252

Tannery, P., 89

Taylor, A. E., 231

Thales, 41, 42, 90, 174

Theophrastus, 40, 42, 73

Theon of Smyrna, 45, 120, 256

Thucydides, 89, 114

Timaeus, 94, 111, 116–37, 177, 216, 218

Time: measurement of, as motive for astronomy, 27, 34, 123; at night, 156

Timocharis, 171

Trigonometry, 23, 155, 221

τροπή, 116, 197, 210

Tropics, 17, 19, 155, 158, 210

'Turnings': of planetary bodies, 18, 32, 59, 78, 94, 116, 130–1, 210

Unmoved mover, 205, 213, 262

Van der Waerden, B. L., 170; on Pythagoreans, 70, 229; on Babylonian astronomy, 173, 251, 254

Venus, 10, 25, 32, 66, 112, 123, 132, 146, 148, 164, 168, 171, 245, 253; in Eudoxan system, 183, 185, 187–8, 256–7; in Callippic system, 191

Webb, E. J., 159, 248, 251

Wedberg, A., 233

Winds, 32

Xenocrates, 243

Xenophanes, 47f., 53

Xenophon, 227

Year: solar, 86, 87, 89, 189; (length of), 87–8, 189, 193; 'perfect', 131; 'great', 75, 88, 130

Zeno, 52

Zodiac: signs of, 17, 87, 156, 254; rising times of, 19, 156; concept of, 113, 115, 119, 172; constellations of, 156, 164

Zones: in Babylonian astronomy, 164, 172; terrestrial, 210, 265